DEVELOPMENTS IN
FOOD PRESERVATION—5

CONTENTS OF VOLUMES 3 AND 4

Volume 3

Volume 4

DEVELOPMENTS IN FOOD PRESERVATION—5

Edited by

STUART THORNE

*Department of Food and Nutritional Sciences,
King's College, University of London, UK*

ELSEVIER APPLIED SCIENCE
LONDON and NEW YORK

ELSEVIER SCIENCE PUBLISHERS LTD
Crown House, Linton Road, Barking, Essex IG11 8JU, England

Sole Distributor in the USA and Canada
ELSEVIER SCIENCE PUBLISHING CO., INC.
655 Avenue of the Americas, New York, NY 10010, USA

WITH 41 TABLES AND 91 ILLUSTRATIONS

© 1989 ELSEVIER SCIENCE PUBLISHERS LTD

British Library Cataloguing in Publication Data

Developments in food preservation
5.
1. Food. Preservation
I. Thorne, Stuart
641.4

ISBN 1-85166-259-6

Library of Congress CIP data applied for

Photoset in Malta by Interprint Ltd.
Printed in Great Britain at the University Press, Cambridge

PREFACE

We have entered a period of consolidation and subtle improvement in food preservation. No completely new processes have been introduced for some time and it seems unlikely that we will see any in the foreseeable future. Instead, attention is turning towards subtle improvements to existing processes, aimed at improvements in product quality and process efficiency. Perhaps most important of all, research is increasingly directed at an improved understanding of the fundamental behaviour of foodstuffs and food processing systems; thermal and physical properties, heat and mass transfer, the behaviour of micro-organisms and deteriorative chemical reactions. An increasing knowledge of all of these and the relationships between them is reducing the empirical content of food preservation operations and effecting a steady improvement in the quality of preserved foods.

Increasing consumer resistance to food additives has forced change on the Food Industry. Whilst we must resist the 'all additives are bad' brigade, the food industry is increasingly considering additives by need as well as legality. The use of additives should be based on ratio of benefit to possible detriment that they effect. Additives are not our primary concern here, but, as fundamental knowledge of processing operations improves, we are able to reduce additive use. For instance, there are many examples in this volume of how careful selection of processing conditions can reduce colour loss in foods; if we can reduce colour loss then we need add less synthetic colouring materials. I do not believe that synthetic additives are inherently less desirable than natural ones, but, because the safety of no food component, either natural or synthetic, can be proved absolutely, the fewer the components the safer the food.

The chapters in this volume represent much recent work on fundamental aspects of food preservation and demonstrate how proper application of this knowledge can result in better product quality and safety and improved process efficiency. And this is in the best interests of both the consumer and ourselves.

A problem with which we have to contend is consumer suspicion; our final customers often do not believe that the food industry always acts in their best interests. This is at least partly due to consumer ignorance. The popular media talk of 'scientific proof' when, of course, science is concerned with disproof and probability. Often 'proof' and 'disproof' are interchangeable; we are almost completely sure of our conclusions. But with food safety this is not so. We can never prove that a food will never harm a single consumer; we can only have proof that it can cause harm. So we have to be honest and admit that 'as far as we can ascertain the product is not harmful'. But this is not enough for our non-scientific customers, who have been led to believe in scientific proof. Better customer relations are needed; we must put aside our arrogance and attempt to explain to the public what we are doing, how we do it and why. Then, perhaps, we will have customers who have faith in us and our products.

Those of us in the European Economic Community have an exciting time ahead. The abolition of all trade barriers within Europe means that we have to have coherent European food legislation by 1991. It will also give us a market of some 300 million customers. But there will be immense problems. Every European nation has workable food legislation that permits production of safe food, but to agree to a single set of legislation is difficult; we all have our own national products and processes, which are not compatible with the requirements of other European nations. The British, for example, allow cereal fillers in sausages and non-dairy fats in ice cream and all of us have different regulations for the use of additives. I am sure that the problems will be solved, but we must ensure that they are solved in a way that permits national variation. As far as food is concerned, we must not all end up as identical Europeans, losing our national identities. Imagine the horrors of the Euroloaf or Eurowine. It may seem far fetched, but bureaucrats are capable of curious behaviour in their enthusiasm for unification. When problems of legislation are solved satisfactorily, the vast European market will be a challenge to the food processing industry. Those who are prepared with quality products will thrive and expand, but I fear that many, smaller firms will fall by the wayside. Unfortunately, it will not always be their own fault if they fail through lack of expertise in the vast new market. And the firms that lose their market to the large international companies will often be the suppliers of the national variations in foodstuffs that we need. The great attraction of a 'Common Market' is that we can be a great political and economic power whilst retaining our national identities and what we eat and drink are a major part of these national identities. Let us all ensure that a tariff-free Europe does not become a Europe of identical eaters and drinkers.

STUART THORNE

CONTENTS

LIST OF CONTRIBUTORS

M. BENNASAR

Centre de Genie et de Technologie Alimentaire, Universite des Sciences et Techniques du Languedoc, Place E. Bataillon, 34060–Montpellier, France

W. CANET

Instituto del Frio (CSIC), Ciudad Universitaria, 28040 Madrid, Spain

C. GUIZARD

Laboratoire de Physico–Chimie Matériaux, E N S C M, 8 Rue de l'ecole Normale, 34075 Montpellier, France

J. R. HEIL

Department of Food Science and Technology, University of California, Davis, California 95616, USA

H. G. KESSLER

Institute for Dairy Science and Food Process Engineering, Technical University Munich, D–8050 Freising–Weihenstephan, FRG

G. M. Rios

Centre de Genie et de Technologie Alimentaire, Universite des Sciences et Techniques du Languedoc, Place E. Bataillon, 34060–Montpellier, France

Sudhir K. Sastry

Department of Agricultural Engineering, The Ohio State University, 590 Woody Hayes Drive, Colombus, Ohio 43210, USA

John N. Sofos

Department of Animal Sciences and Department of Food Science and Human Nutrition, Colorado State University, Fort Collins, Colorado 80523, USA

J. Succar

Research and Development Center, Beatrice/Hunt–Wesson Inc, 1645 West Valencia Drive, Fullerton, California 92633–3899, USA

B. Tarodo de la Fuente

Centre de Genie et de Technologie Alimentaire, Université des Sciences et Techniques du Languedoc, Place E. Bataillon, 34060–Montpellier, France

Stuart Thorne

Department of Food and Nutritional Sciences, Kings's College London (KQC), Campden Hill Road, London W8 7AH, UK

Chapter 1

QUALITY AND STABILITY
OF FROZEN VEGETABLES

W. CANET

*Instituto del Frio (CSIC), Ciudad Universitaria,
Madrid, Spain*

SUMMARY

Freezing is widely acknowledged as the most satisfactory method of preserving vegetables over long periods. This chapter reviews and discusses technological innovations in the field of freezing during the past ten years; primarily on the effects of freezing on the quality of the final product.

Topics considered include basic aspects of freezing itself, the physical and chemical alterations taking place in the product during storage, and the combined effect of storage time and temperature (TTT factors), with a discussion of the relative importance of the freezing rate and storage temperature.

The effect on quality of product–related factors, processing (in particular blanching, in view of the greater importance of technical developments in this area as compared to those relating to freezing per se), pre–freezing treatments, and packaging (PPP factors).

Lastly, future research priorities are summarized. A combination of more basic knowledge, technological development, and optimization of the many PPP and TTT factors would all help achieve the higher levels of final product quality increasingly demanded by consumers.

1

1. INTRODUCTION

Looking back over the historical development of quality requirements for processed foods,[1] freezing is undoubtedly the most satisfactory method for the long-term preservation of vegetable produce. The low temperature commonly recommended for frozen foods ($-18°C$) inhibits the activity of microorganisms and slows the biochemical reactions which, together with the reduction in water activity due to freezing, can maintain the initial quality and nutritive value almost unchanged, with differences only in texture between frozen and fresh vegetable products.

The economic importance and continuous growth of frozen vegetable consumption in Europe are reflected in Table 1. The data show that in 1984 the UK, Federal Republic of Germany, and France were the three countries with the highest overall consumption of frozen foods (poultry products not included), while Sweden, France, and Denmark enjoyed the highest per capita consumption. The increase in consumption in 1984 with respect to 1983 ranged between 5 and 10% in all countries except Sweden (2%) and Holland (3·1%).

The proportion of frozen vegetables (including potato products) in total frozen food consumption ranged from 34% in Sweden and Norway to 61% in Italy. Frozen vegetables were, in volume, by far the largest sector of frozen food sales in all countries. The data on total frozen vegetable consumption in 1985 show the UK in first place, followed by the Federal Republic of Germany and France, with the highest increase in consumption over 1984 being recorded in France (28·9%) and the lowest in Holland (2·8%) and Finland (3%).

The importance of freezing and the scientific and technological development of this method of food preservation is highlighted by the considerable attention afforded to it in all the countries of the EEC in recent years. In 1977, it was included as a research priority in food science and technology under the first Intra-European Collaborative Research Programme; the COST 91 Project on the Effects of Thermal Processing on Quality and Nutritive Value of Foods.

In the three-year period 1981–3, a subgroup for collaborative work on the influence of freezing, distribution, and thawing on the quality and nutritive value of foods existed under this project, culminating in new and interesting scientific findings in the field of freezing in general and with respect to frozen vegetables in particular.

Vegetable produce is characterized by its seasonal and highly perishable nature, with extremely rapid deterioration in quality at ambient temper-

TABLE 1
FROZEN FOOD CONSUMPTION IN EUROPE IN 1984 AND 1985[a]

	Total frozen food consumption[b] (Tm) 1984 A	Per capita consumption (kg) 1984	% difference over 1983	Total frozen vegetables consumption[c] (Tm) 1984 (B)	% frozen vegetables/total frozen foods (B)/(A) 1984	Frozen vegetables[c] consumption (Tm) 1985	% difference over 1984
Denmark	94·769	17	9·1	35·509	37·46	39·355	10·8
Finland	38·511	10·1	5·4	17·754	46·1	18·290	3
France	669·817	18·5	9	313·764	46·8	404·421	28·9
Germany (FR)	692·219	7·5	9·2	390·130	56·3	428·846	9·9
UK	880·000	11·2	5·4	444·000	50·4	493·000	11
Italy	229·030	3·7	10·2	140·640	61·4	154·240	9·6
Holland	205·691	19·3	3·1	109·450	53·2	112·500	2·8
Norway	53·817	11·9	9·7	18·979	34·8	21·777	14·7
Sweden	172·688	20·1	2	59·009	34·2	61·864	4·8
Switzerland	76·982	11·8	7·9	42·065	54·6	46·056	9·5

Tm, metric tons.
[a] Elaborated with values obtained from Elman[2] and Halique.[3]
[b] Frozen poultry products not included.
[c] Including frozen potato products.

atures after harvesting. Freezing under optimum conditions and storage at temperatures of $-18°C$ or lower can maintain initial quality for between twelve and thirty-six months. The final quality of frozen vegetables is dependent upon a large number of factors, such as product type and variety, degree of ripening, initial quality, harvesting methods, elapsed time between harvesting and processing, pre-freezing treatments, freezing procedures, packaging, storage time and temperature. All these factors and their interactions, combined with possible deficiencies in storage conditions during the freezer chain, point out the difficulty in optimizing the process as a whole and thus in satisfying consumer quality expectations.

The main objective of this chapter will be to review the fundamental aspects and technological features of the main stages of the freezing process and storage and their effects on the quality and stability of frozen vegetables, focussing on the state of the art and future research goals.

1.1. Historical Notes on Vegetable Freezing

The modern era of frozen foods began in the 1930s, with the appearance of frozen vegetables in the United States market.[4] Intensive research efforts in the field had commenced during the previous decade and have continued up to the present day, giving rise to an extensive scientific literature on the quality and stability of frozen vegetables. The first review of the literature was made by Tressler and Evers.[5] Initially, research concentrated on the search for solutions to the problems posed by changes in colour and the development of off-flavours during storage. In 1929 Joslyn and Cruess[6] demonstrated that blanching (brief heating of the product) increased the stability of frozen vegetables over prolonged periods of storage. Onslow[7] studied and discussed, for the first time, the phenomena underlying oxidative browning in frozen fruit, relating it to the action of certain oxidases on phenols. In 1941 Joslyn[8] showed that ascorbic acid could act as an inhibitor of the oxidation reactions causing browning.

At the same time, scientists also turned their attention to the effect of the freezing rate on the quality of frozen vegetables. Woodroof[9] described physical changes in vegetable tissues during freezing, showing the advantages of rapid freezing. Morris and Barker[10] confirmed these results for fruits, recording considerable organoleptic differences between fruits subjected to rapid or slow freezing; nevertheless, the differences disappeared after storage. Lee et al.[11,12] were of the opinion that there was no advantage in using faster freezing rates than those commonly employed by the food industry at the time.

In order to maintain product quality and stability after freezing, it was necessary to select a storage temperature appropriate to the probable duration of storage and distribution until consumption. At the end of the 1940s scientific opinion concerning the most suitable storage temperature was still divided, though −18°C (0°F) had become the most widely used temperature industrially; more from technical and cost considerations than on any sound scientific basis. Today −18°C is regarded as the target temperature, but it is not always maintained, particularly during transportation and distribution. At the request of the US frozen food industry, in need of scientific 'facts' on which to base quality preservation during the storage and distribution of frozen products, in 1950 the USDA Western Regional Research Laboratory at Albany, California designed and implemented a research programme aimed at studying the combined action of storage time and storage temperature on the quality and stability of frozen foods, the so-called 'TTT Project' (time, temperature, tolerance). Such concepts as 'high quality life' and 'practical storage life' were first defined by this project, and instrumental and sensory measures of quality were designed and put into practice. The scientific methodology developed under the project is still in use today. The results of this programme were published by van Arsdel et al.[13] Particular attention was addressed to the quality and stability of frozen vegetables, with extensive studies of such vegetables as green beans, peas, spinach, and cauliflower[14] and fruits like strawberries, peaches, raspberries, and cherries.[15,16]

Together with this applied research, more basic theoretical work was performed in an attempt to study and explain the structural changes occurring in different vegetable tissues associated with freezing, thawing, and blanching.[17] More recently, similar work has been carried out, directed specifically at cauliflower,[18] green beans,[19] and peppers.[20] There has been a comprehensive review by Reeve[21] and significant work by Brown[22] on green beans, Rahman et al.[23] on carrots, and Mohr[24] on peas, spinach, and green beans. All these authors underlined the beneficial effect of fast freezing rates on quality by reducing structural damage. Losses in quality that are chemical in nature, caused primarily by enzymatic reactions, gave rise to basic, general papers like those by Joslyn,[25] Lee et al.,[26] Leeson,[27] and a review of blanching by Lee.[28] Interesting work was also carried out by Esslen and Anderson[29] and Zoueil and Esslen[30] for peroxidase and by Sapers and Nickerson[31] for catalase on the degree of thermal destruction of enzymes at high tempratures and possible regeneration during frozen storage. Other work by Dietrich et al.[32] and Pinsent[33] for peas, by Dietrich and Newmann[34] for Brussels sprouts, by Rosoff and Cruess[35] for cauliflower, and by Resende et al.[36] for green beans and spinach was aimed at

determining optimum conditions for blanching and the possible repercussion
of blanching on quality. Aylward and Haisman[37] published an interesting
review on fruit, summarizing the knowledge and results obtained up to that
time with regard to oxidative systems and their effect on the quality of
processed vegetables.

There are other factors in addition to the freezing process itself and to
frozen storage that exert a sizeable influence on quality, e.g. product type
and variety, degree of ripening, initial quality, pre-freezing treatments, and
packaging. Dalhoff and Jul[38] listed these, which they termed PPP factors
(product, processing, packaging), stressing their relative importance—in
some cases more important than freezing rate and storage time and
temperature themselves. Such factors have been reviewed by Gutschmidt,[39]
Olson and Dietrich,[40] and Pointine et al.[41]

Some works dealing with frozen food processing, whose chapters on
vegetables are an excellent compendium of scientific information, are
essential to any list of reference works: Fennema et al.,[42] Cioubanu et al.,[43]
Desrosier and Tressler,[44] and the International Institute of Refrigeration.[45]

This summary of the considerable scientific literature available, of
necessity incomplete, emphasizes topics with a bearing on the quality and
stability of frozen vegetables that were the main areas of interest to research
workers between 1930 and 1970.

In line with the main objective of the present review, attention will focus
on these main subject areas: the effect on quality of the initial freezing
process, the frozen storage period, the TTT factors, and, finally, product-
related factors, processing (in particular blanching), pre-freezing treatments,
and packaging (PPP factors).

2. INFLUENCE OF FREEZING

2.1. The Freezing Process

The freezing process in itself consists of lowering the product temperature
to $-18°C$ at the thermal centre, resulting in crystallization of most of the
water and some solutes.

A knowledge of the temperature changes undergone by the different
parts of the product during freezing (Fig. 1) is essential in order to be able to
understand the process. Three separate stages are distinguishable.[46] First
there is a pre-freezing stage, in which the product is cooled from its initial
temperature to the freezing point. Ice crystallization occurs only after

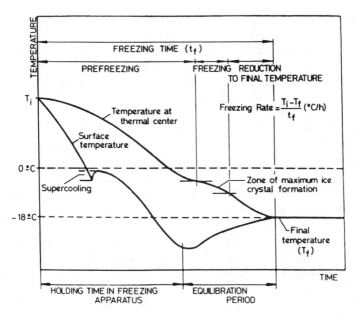

FIG. 1. Product temperature during freezing. Adapted from IIR.[46]

a degree of supercooling, i.e. reduction of the temperature to below the freezing point for some seconds, at temperatures of from -5 to $-9°C$, depending upon the product. Following this, there is a slight rise in the temperature back to the product freezing point, caused by the heat produced by the initial crystallization. In the freezing stage, most of the water in the product undergoes a phase change to ice, with the temperature remaining almost constant due to the concentration of solutes in the still unfrozen aqueous phase, i.e. the zone of maximum ice crystal formation.

The freezing process is not complete until the temperature at the thermal centre has reached the final temperature. Removal of the product from the freezer before this point may result in slow freezing at the thermal centre, with consequent loss of quality. Since the final temperature at the thermal centre should be at least as low as the storage temperature, there is a final stage in which the product is further cooled, from the temperature at which most of the water has been frozen to the storage temperature. The duration of the freezing process depends upon the freezing rate ($°C/h$), defined by the International Institute of Refrigeration[46] to be the difference between initial temperature and final temperature divided by freezing time, freezing time being defined as the time elapsing from the start of the pre-freezing stage until the final temperature has been attained. This will be affected by product size (particularly thickness) and shape, as well as by the parameters of the heat transfer process and the temperature of the cooling medium.

2.2. Ice Crystallization

The pre-freezing stage brings about an increase in the permeability of the membranes in vegetable tissues, with a concomitant loss of intracellular pressure,[47] but the irreversible, adverse effects of freezing on quality are the result of crystallization. The ice formation process comprises crystal nucleation and growth. From a physical standpoint vegetable tissues may be treated as dilute aqueous solutions in which nucleation is heterogeneous, i.e. the non-aqueous particles in the solution act as nuclei around which the water molecules aggregate to form ice crystals. The nucleation rate becomes higher the lower the supercooling temperature, and hence larger numbers of stable crystallization nuclei form as the supercooling temperature falls to below the temperature at which nucleation commences. The rate of crystal growth increases as the cooling rate increases, such that very few nuclei form if the system is held at a temperature between the freezing point and the nucleation temperature. Instead, the nuclei grow larger, giving rise to very big crystals. If the temperature of the system drops very quickly to below the nucleation temperature, large numbers of nuclei form. Crystal growth is then more limited, with crystal size inversely proportional to the number of nuclei.[48]

The temperature at which vegetable tissues start to freeze is directly dependent upon the soluble solids content—in particular sugars, salts, and acids—rather than on the water content. As the product progressively cools to below the initial freezing point, more and more water turns into ice, so that the residual solution becomes more and more concentrated. The amount of ice that forms at a given temperature is also related to the initial soluble solids content.[49] Once the water has begun to freeze, the crystallization rate and the location of the crystals depends upon the rate of heat removal (freezing rate), tissue structure, and the rate of diffusion of water from the solutions to the surface of the ice crystals. At low freezing rates, few crystallization nuclei form, and the cells lose water by extracellular diffusion through the cell membranes, leading to the concentration of solutes that impedes the formation of crystallization nuclei inside the cells. Consequently, the extracellular crystals grow to a relatively large size, resulting in progressive separation of the cells and plasmolysis of the cell protoplasm, which can bring about partial to total collapse, depending upon the rigidity of the cell structure. Conversely, when the tissues are frozen at a rapid rate, to a low enough temperature at the thermal centre, numerous, tiny ice crystals are distributed evenly inside and outside the cells. Rapid freezing brings about minimal migration of water to the crystallization sites and thus only minor modifications in tissue structure.

2.3. Effects of Freezing on Structure and Texture

A considerable amount of research has been carried out on the effects of ice crystal size and location on the structural and textural quality of frozen vegetables. These two factors, together with the increase in volume that occurs during freezing, result in more or less irreversible damage to the structure. The intensity of such damage depends upon the freezing rate and the product structure itself. Brown[50] showed the differing effect of freezing rate (for freezing times of 1, 5, and 20 or more min) on tissue structure in asparagus and green beans. Ice crystals grow larger at slower freezing rates, causing more damage to the cell structure. Structural deterioration was more pronounced in green beans, which have a parenchymal structure consisting of larger cells and more vacuoles, than in asparagus, in which the structure is mainly protoplasmic, with a relatively low level of vacuolation. These alterations in tissue structure during freezing bring about an irreversible deterioration of product texture.[21,51-54] The beneficial effects of rapid freezing rates on structure and texture have been reflected by the results of various methods of texture analysis (i.e. histological, sensory, imitative, and objective techniques) in studies by Gutschmidt[49] on green beans, Brown[22] on green beans, Mohr[24] on green beans, spinach, and peas, Canet et al.[55,56] on potato, and Canet and Espinosa[57-59] on carrots and peas. Monzini et al.[60] used histological examinations to evaluate the beneficial effect of rapid freezing rates, demonstrating the adverse effect of blanching and cooking, which mask the different structural alterations caused by rapid or slow freezing. In contrast, other studies using objective methods of texture analysis have successfully detected the effect of rapid freezing rates, even after cooking, in green beans and carrots,[61] potato and carrots,[55,56] and in carrots and peas.[57,59]

Fruits are particularly sensitive to changes in texture. Freezing results in the loss of internal cell pressure (turgor), with cell walls rupturing at slow freezing rates, giving rise to softening of the tissues and the release of large amounts of exudate during thawing. Rapid freezing using liquid nitrogen (LN_2) or liquid fluorocarbon freezant (LFF) together with gradual thawing improved fragility and texture and reduced the amount of exudate.[49,52]

2.4. Importance of the Freezing Rate

The close correlation assumed between frozen food quality and rapid freezing rates is widely accepted, despite a large body of evidence suggesting that fast freezing rates do not always result in especially high frozen food quality. The persistence of this assumed relationship was discussed at length by Jul,[62] who explained its longevity by its ready understandability; it has the support of manufactures of freezing equipment and associations of frozen food producers, who maintain that faster freezing rates yield superior end products.

EEC countries have many different laws, regulations, standards, and codes of practice governing foods and food processing; at present they require deep frozen foods to be frozen very rapidly, within specified short times. The importance of the effect of freezing rates on frozen vegetable quality has also been studied by Delaunay and Rosset[63] and by Ulrich.[64]

The accepted influence in vegetable freezing of other thermal treatments such as blanching, thawing and cooking, and of other factors such as product type, packaging, and storage time, have reduced the importance once attached to the freezing rate as the prime factor affecting final quality. This is not meant to suggest that the freezing rate does not in fact affect quality, since it is true that most fruits and vegetables lose quality when frozen slowly. As previously pointed out, strawberries, green beans, carrots, potatoes, mushrooms, and corn-on-the-cob all exhibit improved texture and higher water retention when frozen rapidly. Other products, such as peas, are less sensitive to freezing rate, and only extremely fragile foods like tomatoes derive any real benefit from rapid freezing (faster than 10 cm/h).

The recommended freezing rates of more than 0·5 cm/h for packaged frozen foods and of from 5–10 cm/h for individually frozen products (IQF = individual quick freezing) are considered adequate to attain high-quality frozen products.[45]

The freezing rates commonly achieved today using commercially available freezing equipment are such that the importance of the freezing rate in the freezing process as a whole has been reduced; and, in practice, when selecting the freezing method to be applied, other technical and economic factors, such as control and monitoring, processing capacity, energy savings, processing times, and capital and operating costs are taken into account, in addition to product quality characteristics (texture, colour, dehydration, exudate, etc.).[65]

3. INFLUENCE OF STORAGE

Frozen foods do not remain stable over the storage period, which frequently lasts for months or even years, but lose quality, the extent of which depends upon the storage temperature and product type. The quality losses in frozen vegetables stored at −18°C can be greater than those sustained during initial freezing, pre-freezing treatments, or thawing. They are caused solely by physical and chemical alterations taking place within the product itself, inasmuch as microbial growth is not a factor at the temperatures involved.

3.1. Physical Changes Occurring during Storage

Physical changes affecting frozen vegetable quality during the storage period include what are termed recrystallization and sublimation phenomena, related to the stability of the ice crystals inside and on the surface of the product. Recrystallization comprises changes in the number, size, shape, and orientation of the ice crystals that form after initial solidification during freezing and is the result of the successive melting on the surface of small ice crystals, followed by recrystallization on the surface of larger crystals, thereby increasing the size of these while reducing the total number of crystals. The theory of recrystallization has been discussed at length by Calvelo.[66] This phenomenon has been recorded in a wide range of foods, including vegetable tissues. Fruits frozen both rapidly and slowly and stored at −18°C for six months were found to have the same size and distribution of ice crystals.[12] The effect of recrystallization during storage and distribution can eliminate benefits derived from rapid freezing.[62] Figure 2 shows the variation in ice crystal length in tissue frozen at different freezing times (B) and after three months in storage at −20°C (A).[67] Crystal growth of this kind is known to be greatly accelerated at higher storage temperatures and by fluctuations in temperature.

Monzini et al.[60] showed that the pre-existing tissue and cell alterations had not basically deteriorated after six months of storage at −20°C (±1°C). This indicates that recrystallization does not take place extensively at low temperatures over average storage periods. Fennema et al.[42] and Zaritzky et al.[68] found crystal growth at warm temperatures but not below −10°C. Sublimation of ice at the surface can also occur during storage in inadequately packaged food, leading to desiccation, with accumulation of the water thus extracted in the form of frost inside the packaging. In addition to causing undesirable weight loss, excessive desiccation can speed

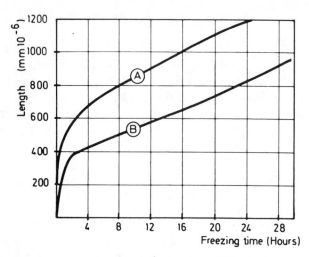

FIG. 2. Crystal size in relation to the freezing time. Immediately after freezing (B) and after three months in storage at $-20°C$(A). After Aström.[67]

up oxidative alterations at the surface of the product, adversely affecting quality.

Both recrystallization and surface desiccation are accelerated by fluctuations in storage temperature, although the importance of this declines at low storage temperatures.[45] There is no evidence that temperature fluctuations at temperatures below $-18°C$ lead to reductions in frozen food quality.[64] The contradictory nature of the few results available on the effect of recrystallization and sublimation on quality call for further in-depth studies of these phenomena.[69]

3.2. Chemical Changes during Storage
Despite the low temperatures, a series of enzymatic and non-enzymatic chemical reactions during frozen storage takes place in vegetables. Their influence on quality is particularly important, because these reactions are associated with the appearance of off-flavours and -odours, changes in colour due to the breakdown of the chlorophylls and other natural pigments, and the development of enzymatic browning, as well as the autoxidation of ascorbic acid. Changes in pH during freezing and frozen storage may also be related to alterations in the kinetics of such reactions and the decrease in quality.[42]

3.2.1. Changes Associated with the Appearance of Off-Flavours and -Odours
During frozen storage, ethanol (produced by glycolysis) and other volatile compounds may accumulate in the tissues of vegetables that have not been adequately blanched. Such accumulation coincides with the development of off-flavours and -odours that can persist even after cooking.[70,71] Such off-flavours and -odours are considered to be at least partially a consequence of the enzymatic oxidation of lipids.[37] Studies on unblanched peas have related the development of off-flavours and -odours with increases in the acidity and peroxide content and with the accumulation of volatile substances (hexanal and other aldehydes) produced by oxidation of polyunsaturated fatty acids by lipoxygenase.[72] Higher concentrations of hexanal have been recorded in peas damaged during harvesting. These derivatives of lipid hydrolysis and the lipoxygenase-induced oxidation of unsaturated fatty acids do not appear to be the only or even the main source of alterations in odour and flavour.[73] When unblanched peas were packaged in a nitrogen atmosphere to prevent oxidation reactions during storage, the same off-flavours were found to develop as when the peas were packaged in a normal atmosphere.[72] A wide variety of related substances, such as ethanol, acetaldehyde, large amounts of hexanal, several alcohols, and other volatile organic compounds, have been identified by chromatography, but it has not been possible to establish any direct relationship between any individual substances and the characteristic hay and rancid tastes in inadequately blanched vegetables.[74] Blanching for a sufficiently long time destroys the oxidases present, and off-flavours and -odours do not develop during storage.[75] However the origins of such flavour and odour alterations have not yet been established and indicate the need for research into the chemical and biochemical mechanisms involved.[76,77]

3.2.2. Changes Associated with Alterations in Colour
Frozen vegetables undergo more or less intense alterations in colour during storage, brought about by changes in the natural pigments, chlorophylls, anthocyanins, and carotenoids or by enzymatic browning. The characteristic green colour of frozen peas, green beans, and spinach gradually turns brown during storage at $-18°C$, due to the transformation of chlorophyll a and chlorophyll b into their corresponding pheophytins. Such changes in the initial colour occur much more quickly in unblanched vegetables or those stored at insufficiently low temperatures. A substitution reaction replacing the Mg^{2+} ions in the porphyrin nucleus with protons is responsible for these changes. This reaction is favoured by the natural acidity of vegetable tissues and takes place even in the absence of enzyme

activity or oxygen. Another important path of chlorophyll breakdown is via the action of the peroxides produced by the lipoxygenase-induced oxidation of polyunsaturated fatty acids in the presence of oxygen.[78] These latter reactions proceed slowly by purely chemical mechanisms,[79] which is why the chlorophylls are more stable over storage in blanched vegetables.[80] The influence of such alterations in colour during storage on final product quality is considerable, since differences in chlorophyll content of as little as 4% in green beans and 1·5% in peas have been detected by 85–95% of taste panellists.[81] Blanching for a very brief time at high temperature is less detrimental to pigments than blanching for a longer time at lower temperatures,[32] which can speed up pigment degradation during storage.[79] Such pigment degradation, and the effect of blanching on it, are to a large extent dependent upon the type and variety of vegetable involved, along with its degree of ripening and tissue damage sustained during harvesting.[42] Table 2 shows the rate of chlorophyll degradation at various storage temperatures. The rate can be seen to be higher in green beans and chopped spinach, in which increased processing led to more rapid breakdown of chlorophyll during storage than in the peas or whole leaf spinach.

TABLE 2

FROZEN STORAGE TIME (MONTHS) REQUIRED FOR 10% DECREASE IN CHLOROPHYLL CONTENT OF SELECTED GREEN VEGETABLES

Product	Storage temperature		
	$-18°C$	$-12°C$	$-7°C$
Peas	43	12	2·5
Spinach, leaf	30	6	1·6
Spinach, chopped	14	3	0·7
Green beans	10	3	0·7

Source: Olson and Dietrich.[40]

Anthocyanins are water-soluble pigments responsible for the red colours in fruits and vegetables, and under certain conditions they can be destroyed by enzyme-induced oxidation of the polyphenols. In order to prevent this type of colour loss during processing and storage, blanching is carried out in a sugar solution. Although this has an undesirable effect on texture, it is common practice in red fruits rich in polyphenols and polyphenol oxidase that are intended for use in the manufacture of other products.[80]

The oxidation of carotenoids, liposoluble pigments abundant in many vegetables in the form of xanthophylls, carotenes, and lycopene isomers, is only a secondary cause of alterations in colour, but it is worthwhile to prevent any substantial degradation of these pigments in view of their role in protecting the chlorophylls against oxidation[82] and their function as provitamin A. Blanching protects carotenoids from oxidation by lipoxygenase and by the peroxides derived from polyunsaturated fatty acids.[82]

Alterations in colour during the storage of frozen vegetables as a result of enzymatic browning are caused by the oxidation of phenols in the presence of oxygen in such products as apples, peaches, pears, cherries, cauliflower, potatoes, and mushrooms. The reaction is catalysed by polyphenol oxidases, giving rise to quinones that condense in the form of brown-coloured or reddish brown compounds with a more or less well-defined chemical composition. The quinones in turn act as oxidants for other substrates such as ascorbic acid, anthocyanins, etc. Such enzymatic browning can be minimized by thermal inactivation of the enzymes, the addition of inhibitors, or the exclusion of oxygen.[48] Blanching is clearly the most appropriate method of preventing enzymatic browning. Enzyme activity can also be controlled in mushrooms by adding citric acid (1·5%) to the blanching water or in cauliflower by adding metabisulphite (2 g/litre) to the cooling water after blanching. Adding sequestering agents, e.g. disodium dihydrogen pyrophosphate to the blanching water prevented the subsequent appearance of undesirable colours in potato products.[80,83]

3.2.3. Ascorbic Acid Oxidation and Changes in pH

In inadequately blanched vegetable tissues, ascorbate oxidase catalyses oxidation of ascorbic acid during storage, and this process is accelerated when oxygen-permeable packaging is employed. The oxidation rate depends upon storage temperature and product pH. Figure 3 graphically represents the effect of storage temperature on the ascorbic acid content in spinach. After storage for one year at $-18°C$, the loss was barely 10%, but losses rose swiftly as the temperature increased to $-12°C$ or higher. Losses were similar in peas but higher in green beans and cauliflower. Table 3 presents the times required for 10% and 50% decreases in the initial ascorbic acid content in several frozen fruits and vegetables stored at $-18°C$. Temperature increases of just 2·3°C were extremely important in the case of peas, strawberries, and green beans, halving the storage time required to achieve a 50% decrease in the initial ascorbic acid content.[64] Depending upon concentration, ascorbic acid acts as a reducing agent, maintaining the polyphenols in a reduced state. In tissues that are

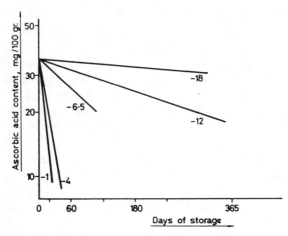

FIG. 3. Effect of storage temperature on spinach ascorbic acid content. After Dietrich et al.[84] Reproduced with permission of The Institute of Food Technologists.

unblanched, and hence in which polyphenol oxidases are active, the ascorbic acid is gradually oxidized, under adverse storage conditions leading to the progressive loss of ascorbic acid and to tissue decoloration. The stability of ascorbic acid rises as the pH drops. In strawberries, which have a low pH, the ascorbic acid is oxidized at a slower rate than in most green vegetables stored under similar conditions.[42] During freezing organic and inorganic salts are concentrated in the aqueous phase. As the temperature falls, the salts precipitate at their respective eutectic points, and as a consequence the pH of the non-freezable aqueous phase changes.

TABLE 3

STORAGE TIMES TO 10% (D_{10}) AND 50% (D_{50}) REDUCTIONS IN THE INITIAL ASCORBIC ACID CONTENT AT $-18°C$ AND TEMPERATURE DIFFERENCE REQUIRED TO HALVE THE TIME TO $D_{50}(T_{50})$

Product	$D_{50}\ (-18°C)$ (months)	$D_{10}\ (-18°C)$ (months)	$T_{50}\ (°C)$
Peas	70	10·6	2·3
Strawberries	30	4·6	2·3
Green beans	30	4·6	2·3
Cauliflower	10	1·5	12·0

Source: Polensky.[85]

Figure 4 illustrates the changes in pH in frozen cauliflower stored at − 10°C and at − 18°C. The pH dropped after freezing and then underwent inflection, the intensity and direction of which varied with the temperature and product. The behaviour of peas and green beans was similar to that of cauliflower.[86] Van den Berg[87] suggested that the decrease in the initial pH

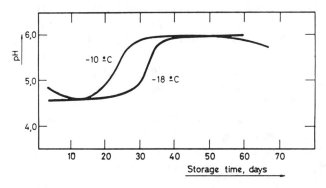

FIG. 4. Changes in pH of minced cauliflower stored at − 10°C and − 18°C. After van den Berg.[86]

subsequent to freezing was due to precipitation of the alkaline phosphates of calcium, magnesium, and sodium. Precipitation of the acid phosphates of potassium and sodium, and potassium citrate was held to be the cause of the final increase in the pH. It is clear that the changes in the pH may play a role in the altered activity of certain enzymes. However, no studies have been able to relate changes in pH to lowered quality during storage.[42] Work in this area would therefore be of interest.

3.3. Combined Effect of Time and Temperature during Storage (TTT Factors)

Deterioration of initial quality due to the physical and chemical changes undergone by products during storage is a function of storage temperature and duration. The combined effect of these two factors, time and temperature (TT), determine product tolerance (T) to frozen storage. Until the development of what are known as the Albany Time–Temperature–Tolerance tests at the USDA Western Regional Research Center at Berkeley,[88] studies on the influence of frozen storage on foods were characterized by intensive but unorganized reporting of data. In the Albany

TTT Project, a large number of fruits and vegetables were tested at various freezer storage temperatures for various periods of time. Quality was measured by objective tests of ascorbic acid degradation, transformation of chlorophyll into pheophytin, and alterations in colour, and by organoleptic tests. The latter usually took the form of triangle tests using a taste panel, indicating when changes in quality were recorded. High quality over storage, or high-quality life (HQL), was defined as the storage period during which initial quality was maintained, from the time of freezing to when 70% of taste panel members were capable of distinguishing differences between foods stored at various temperatures and controls stored at $-40°C$.[15]

Jul[62] and the IIR[46] referred to these keeping times in this type of experiment as the time of a 'just noticeable difference (JND)' or as 'stability time'. Stability times during which initial quality is maintained in storage show an exponential relationship with decreasing temperature. Representing time on a semi-logarithmic scale against temperature yields linear relationships for mean storage time to the first JND in quality on temperature in the range of from $-10°C$ to $-30°C$ similar to those shown in Fig. 5 for colour and taste in cauliflower. Publications from the TTT Project have been the source of extremely valuable information on the stability of frozen vegetables, in particular green beans, peas, spinach, and cauliflower. Table 4 gives the days in storage at several temperatures to detection of changes in colour or flavour. For any given temperature, colour can be observed to be the limiting attribute, as colour changes were detected before flavour changes in all the vegetables tested except spinach. The time prior to a noticeable change in colour quality increased at colder temperatures (range: -4 to $-18°C$).[14]

When a triangle test is carried out by an experienced taste panel, only minor differences are required for detection of a change in quality. It is therefore highly likely that the stability times, or periods of high-quality life, recorded do not represent finite storage times after which products cannot be consumed, but rather the length of time during which frozen products retain their initial quality levels.[15] Most products remain commercially acceptable even after the end of the stability time, and the terms practical storage life (PSL) or acceptability time[62] are used to designate the storage period during which product quality stays at a level acceptable for consumption or for use in further processing.[46] No precise relationship between HQL and PSL has been demonstrated. The collective assessments of scientists, experienced industry representatives, consumers, and taste panellists indicate that for most foods the PSL is from two to five times longer than the HQL. However, in colour-sensitive products like cauliflower,

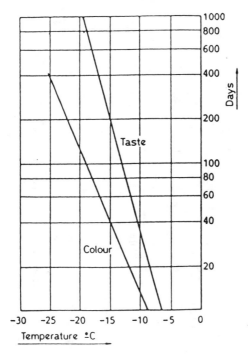

FIG. 5. Stability times for cauliflower determined in the Albany test series. After Dietrich et al.[89]

the PSL may be only slightly longer than the HQL.[46] Table 5 reflects the practical storage life or acceptability times for most frozen fruits and vegetables.[46] The mean PSL's or acceptability times indicated by the IIR[46] at three different storage temperatures were based on products with high initial quality levels that had been properly processed and packaged. Using lower quality raw material processed or packaged inadequately and/or fluctuations in the storage temperature can substantially shorten the storage period in which quality remains acceptable.

The values appearing in Table 5 should therefore not be considered absolute limits to be applied rigidly. In most circumstances, and in accordance with IIR criteria,[46] most vegetables can be categorized as high stables (with a PSL of 15 months at −18°C) or medium stables (with a PSL of from 8–15 months at −18°C), with only peppers being considered low stables (with a PSL of less than 8 months at −18°C).

TABLE 4
DAYS IN STORAGE AT VARIOUS TEMPERATURES REQUIRED TO BRING ABOUT A PERCEPTIBLE CHANGE IN QUALITY

°C	Beans		Peas		Cauliflower		Spinach	
	Colour	Flavour	Colour	Flavour	Colour	Flavour	Colour	Flavour
−18	101	296	202	305	58	291	350	150
−12	28	94	48	90	18	61	70	60
−9·5	15	53	23	49	10	28	35	30
−7	8	30	11	27	6	13	20	20
−4	4	17	5	14	3	6	7	8

Source: Olson and Dietrich.[14] Reproduced with permission of The Institute of Food Technologists.

TABLE 5
PRACTICAL STORAGE LIFE (*PSL*) IN MONTHS AT SEVERAL STORAGE TEMPERATURES

Product	$-12°C$ ($10°F$)	$-18°C$ ($0°F$)	$-24°C$ ($-12°F$)
Fruits			
Raspberries/strawberries (raw)	5	24	>24
Raspberries/strawberries in sugar	3	24	>24
Peaches, apricots, cherries (raw)	4	18	>24
Peaches, apricots, cherries in sugar	3	18	>24
Fruit juice concentrate	—	24	>24
Vegetables			
Asparagus (with green spears)	3	12	>24
Beans, green	4	15	>24
Beans, Lima	—	18	>24
Broccoli	—	15	24
Brussels sprouts	6	15	>24
Carrots	10	18	>24
Cauliflower	4	12	24
Corn-on-the-cob	—	12	18
Cut corn	4	15	>24
Mushrooms (cultivated)	2	8	>24
Peas, green	6	24	>24
Peppers, red and green	—	6	12
Potatoes, French fried	9	24	>24
Spinach (chopped)	4	18	>24
Onions	—	10	15
Leeks (blanched)	—	18	—

Source: IIR.[46]

Once the relationship between temperature and the storage time to a given reduction in quality levels (acceptability times) for a given product has been established, and bearing in mind that the combined effect of time and temperature are cumulative and irreversible over the storage period and that the sequence of events does not affect the total cumulative loss in quality, the total reduction in practical storage life of that product during storage and distribution can be estimated. Figure 6 depicts the acceptability time curve for peas at various storage temperatures. Using this relationship, if the residence times and temperature at each component, or link, in the freezer chain are known, the partial percentage loss of acceptability and total loss of PSL can be calculated, as shown in Table 6. In the table (using peas), the loss was 70% over a storage period of 344 days;[62] when the loss

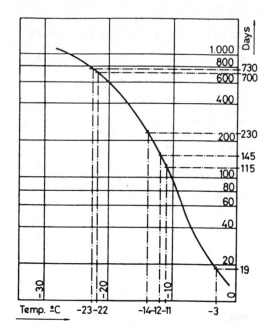

FIG. 6. Acceptability time for frozen peas, based on Gutschmidt, quoted by Spiess.[90]

TABLE 6
LOSS OF PRACTICAL STORAGE LIFE IN FROZEN PEAS

Stage	Time (days)	Temperature (°C)	Acceptability (days)	Loss per day	Loss (%)
Producer	250	−22	700	0·001 429	34
Transport	2	−14	230	0·004 348	0
Wholesaler	50	−23	730	0·001 370	6
Transport	1	−12	145	0·006 667	0
Retailer	21	−11	115	0·008 696	18
Transport	0·1	− 3	19	0·052 632	1
Subtotal	324				59
Home freezer	20	−13	204	0·054 05	11
Total loss of PSL	344				70

Source: Jul.[62]

reaches 100%, the product is no longer suitable for consumption. Similar losses in acceptability were recorded for strawberries,[62] for which the loss amounted to 77% after storage for 344 days. These relatively high figures are explicable given that they are based on old TTT data,[14] and thus fail to take into account recent improvements in product selection, processing, and packaging. Lower total losses of PSL were reported by Muñoz-Delgado[91] for peas and cauliflower, the losses amounting to 37% and 57%, respectively, after storage and distribution during 257 days, although the residence times at the different links in the freezer chain differed and more recent acceptability time values were applied.

Applying the same processing procedures to two different varieties of green beans grown in two separate countries (Germany and Holland), Steinbuch et al.[92] obtained stability results which suggested that it is not possible to draw any general conclusions concerning behaviour during storage, because of the marked effect of vegetable variety and differing consumer preferences. Moreover, differences in the selection of quality attributes, the relative importance attached to each of the various attributes by consumers in different countries, and the differing objective and subjective methods used to measure such attributes are all factors exerting a significant influence on the substantial variability in the results that have been reported in the literature with respect to PSL and HQL. A TTT curve is only valid for a given product with a given raw material quality that has undergone a given process in a given type of packaging. These so-called PPP (product, process, packaging) factors, together with the taste panel factor, can be as decisive to product quality and stability as storage temperature and time.

There is a clear need to investigate the importance of the different TTT and PPP factors on the stability of frozen vegetables. Tests to determine the acceptability of products with known levels of initial quality and tolerance to storage at each of the different links in the freezer chain are also required, in order to be able to assess the true effect and relative importance of each of the links on final quality.

3.4. The Importance of Storage Temperature in the Freezer Chain

The effect of storage temperature on product stability is expressed in TTT curves by the Q_{10} factor, which indicates the proportion by which product quality retention time is augmented for every 10°C decrease in temperature. In frozen vegetables stored at between -15 and $-25°C$, this factor is quite variable between the values of 3 and 6, and it is clear that temperature is a much more important factor in determining quality loss than is storage time. As a result, knowing the exact amount of time that has elapsed from freezing or packaging is not overly important; these dates can mislead consumers with respect to product quality if the product has not been stored at a sufficiently low and constant temperature. Under certain conditions losses in quality are higher and occur faster than might otherwise be expected on the basis of the cumulative effect of storage time and temperature. This is true in the case of fluctuations in storage temperature. Various authors have attached different degrees of importance to storage temperature fluctuations. In a thorough review of the question, Ulrich[64] concluded that, with the exception of certain special cases, fluctuations in storage temperature had no particular effect on product quality. Conversely, van den Berg,[93] in tests with individually frozen (IQF) vegetable products sealed in plastic packaging and exposed to radiation while on sale in freezer display cabinets, detected alterations in colour and flavour, together with weight losses ranging from 2–8%, in peas that had undergone temperature fluctuations. Similarly, Hawkins et al.,[94] also working with peas, reported severe colour and ice formation defects due to temperature fluctuations; these defects were in large part avoided when the freezer display cabinets were covered overnight.

The importance of storage temperature fluctuations in the freezer chain and the influence of improper handling are thus also areas in need of attention by scientists in order to determine their actual effect on product quality. The recommendation for the lowest possible storage temperatures ($-30°C$) during production and at wholesalers and for $-18°C$ in sales cases and consumers' freezers has been modified in recent years out of energy considerations. Ulrich[64] was of the opinion that it should be possible to achieve adequate levels of quality at storage temperatures higher than $-18°C$, providing that the temperature did not rise too high for too long. Jul[62] recommended $-22°C$ during storage at wholesalers, together with reasonable precautions to avoid temperature increases during transportation, with temperatures lower than $-10°C$ at all points below the load line in retail display freezers. The IIR[46] considered the commonly accepted temperature of $-18°C$ to be the upper limit for storing most vegetables from one season to the next while allowing for a reasonable overlap. Table 5 indicates that in practice the PSL for most vegetables was substantially in excess of one year, provided that the packaging material used afforded adequate protection against moisture migrations and temperature fluctuations. Only mushrooms and asparagus needed temperatures of $-25°C$ or colder to attain a storage life of one year. Aromatic herbs such as parsley, chives, and basil, when unblanched, required $-30°C$, and even then the storage life was less than a year.

As can be seen from Table 6, production and wholesale storage temperatures should be as low as possible (-25 to $-30°C$) and as constant as possible, since this is where most products spend the greater part of their storage lives between freezing and final consumption. Transportation is normally very short, and its influence on quality is, in most cases, negligible. On the other hand, the time in the retail display cabinet and home freezer is critical, because the temperatures there tend to be warmer and more subject to fluctuations.

4. INFLUENCE OF PPP (PRODUCT–PROCESS–PACKAGING) FACTORS

The product, processing, and packaging factors briefly referred to in previous sections constitute an important field of research, because of their marked influence on frozen vegetable quality and stability during storage and distribution.

4.1. The Product

The raw material used in the preparation of frozen vegetables is an important conditioning factor affecting both the quality and nutritive value of the final product. A product's suitability for freezing is determined by agrotechnical practices and conditions, the species and variety involved, the degree of ripening, and the time elapsing between harvesting and processing. It is thus obvious that only raw material that is clean, sound, and of high quality should be selected for freezing. Agrotechnical factors include soil type, climatic conditions, and excessive rainfall or irrigation, all of which exert a pronounced effect on fresh product quality and on product deterioration during processing or cooking prior to consumption. Fertilizer composition and fertilizer application conditions also affect product suitability for freezing. Large amounts of nitrogen fertilizers have a beneficial effect on fruits like peaches, enhancing the texture, but a negative effect on spinach, due to the accumulation in the leaves of nitrates that may turn into toxic nitrites. The application of growth regulators at the appropriate times may appreciably reduce susceptibility to browning in fruits.[95] The textural attributes of frozen potato products may be improved by cultural practices designed to augment the solids content. Field-grown vegetables tend to maintain better texture after freezing, while products grown using forced cultivation methods tend to be unsuitable for freezing.[46]

Most vegetable products can be frozen, and excellent results have been achieved using green beans, spinach, peas, cauliflower, Brussels sprouts, sweet corn, blueberries, raspberries, strawberries, etc. The potato deserves to be singled out in view of its excellent freezing qualities, which have in turn led to the appearance of a large number of commercial frozen potato products. In contrast, the flavour and texture of certain other produce, such as bananas, grapes, tomatoes, and lettuce, undergo considerable alteration when subjected to freezing. The variety employed is usually of fundamental importance in order to minimize such alterations. Extensive research has been carried out for the purpose of selecting and breeding varieties suitable for freezing that will deliver high levels of quality in terms of texture and flavour in the final product.[96-102] Research workers in the field should turn greater attention to this aspect, with the aim of expanding the number of available varieties suitable for mechanical harvesting and handling that afford high yields and optimum levels of quality both as raw material and as the final frozen product.

The complete cessation of physiological processes (tissue death) brought about by blanching or by freezing itself requires that harvesting of fruits and vegetables to be frozen should be carried out at just the right moment of ripening. This, together with the commercial need for continuous supplies of raw material, means that the proper organization of varieties used, planting times, and growing zones is essential, as are proper harvesting, transportation, and handling tailored to each individual product, which may otherwise quickly lose their nutritive value after harvesting. The higher the temperatures and the longer the time that elapses between harvesting and processing, the greater the losses. Spinach loses 75% of its initial ascorbic acid content within 24 h at ambient temperature; in 18 h green beans lose 8% of their initial ascorbic acid at 4°C and 30% at 20°C, with losses climbing to 75% of initial levels in 42 h at 20°C. By way of comparison, the combined action of blanching and freezing results in a loss of only 10%.

Between harvesting and processing, products become dehydrated, undergo wilting to a greater or lesser extent, and lose their shiny, turgid appearance. Asparagus turns yellow and fibrous if not rapidly submerged in iced water; mushrooms, when damaged, undergo intense browning if not processed very quickly; peas produce considerable respiratory heat accompanied by the development of off-flavours and a substantial rise in tenderometer values; strawberries undergo a loss in quality in 11 h at 21°C equivalent to that sustained over an entire year when stored frozen at −18°C.[46,95]

Pre-cooling and low-temperature transportation and storage retard such post-harvesting losses. Research into the specific mechanisms producing these quality losses in individual products is needed, as in the selection and breeding of varieties resistant to and methods for minimizing such losses when prolonged periods between harvesting and processing are unavoidable.[77] To cite just one example, there is still a need to obtain fruit varieties with low levels of polyphenol oxidase or oxidizable phenols which are less susceptible to browning before freezing or during frozen storage.[103]

4.2. Processing

4.2.1. Preservation and Pre-freezing Treatments

The purpose of preparation and protective pre-freezing treatments is to obtain a product ready for consumption, giving it optimum conditions for withstanding storage. The most common preparatory procedures are grading, selection, washing and peeling; also, specifically for fruits, destoning and slicing; specifically for vegetables, shelling, trimming, chopping, and slicing.[104] The most common protective treatments include the blanching of vegetables, the addition of enzymatic browning inhibitors (ascorbic acid, sulphur dioxide, citric acid, sodium chloride), and, for fruits, the removal of free oxygen by sweetening or proper packaging.[95]

4.2.1.1. Main preparatory procedures. Washing cleans the product of dirt and impurities (soil and waste matter), of pesticide residues, and, in the case of vegetables, of up to 90% of the microbial flora. Lightly chlorinating the wash water by adding gaseous chlorine or sodium hypochlorite enhances the action of the water, preventing the formation of sludges of bacterial origin in the equipment and the development of unpleasant odours. Free chlorine contents of the order of 5–10 ppm have no adverse effect on product flavour nor any corrosive action on equipment. Washing accounts for a high proportion of the total water expended in the process (25–35 litres per kg of frozen fruit or vegetable).

Peeling, one of the most delicate pre-treatments, is performed industrially by abrasion, high-pressure steam, treatment with sodium hydroxide solution or mechanically. Abrasion is effected by removing the outer surface of the product by abrasion with rough, moving surfaces, but it has the drawback of considerable loss of raw material. Steam peeling is carried out by heating the product to a temperature of up to 80°C and subjecting it to pressures of from 4–7 kg/cm^2 for between 30 seconds and 3 minutes. Peeling by sodium hydroxide solution is also carried out after a pre-heating stage followed by immersion of the product in a 10–20% solution at a temperature of from 60–90°C for between 1 and 5 minutes, according to product type.[46,95] The disadvantage of all these methods is the substantial losses of raw material involved (8–20% in potatoes, depending upon their shape and age). Using sodium hydroxide with infra-red heating enabled Graham[105] to lower sodium hydroxide solution consumption by 80%, decrease raw material losses by one-third, and reduce water usage by up to 95%. Abrasion is commonly employed for potatoes and carrots; chemical peeling is utilized for fruits, especially peaches; mechanical peeling is used for pears and apples.[104]

After washing and peeling, the product may, depending upon product type and variety, be subjected to procedures such as sorting, paring, stemming, trimming, cutting, or pulping. All these operations must be carried out with the utmost care under the most stringent hygienic conditions in order to prevent contamination of the product and mechanical damage. The varying degrees of complexity and automation of the process according to product type require a thorough understanding of the mechanical properties of each individual product, and further progress in this field, with a view to improving automation and optimizing procedures, is desirable.[54]

4.2.1.2. Blanching.

Blanching is a thermal treatment commonly applied in a variety of vegetable preservation treatments and is particularly important in freezing because of its very considerable influence on quality. The product is heated, typically by brief immersion in water, to a temperature of from 85–100°C, or in steam, to a temperature of 100°C. The primary objective is to inactivate enzymes responsible for alterations in the sensory quality attributes (off-flavours and -odours) and in the nutritive value (vitamin losses) taking place during storage. Blanching also affords a series of secondary benefits, in that it destroys vegetative cells of microorganisms present on the surface, complementing the action of washing, eliminates any remaining insecticide residues, enhances the colour of green vegetables, and eliminates off-flavours produced by gases and other volatile substances that may have formed in the time between harvesting and processing. The duration of blanching varies according to the method employed, product type and variety, product size, and degree of ripening, but the chief factor affecting processing time is the blanch temperature. Oxidases, peroxidases, catalases, and lipoxygenases are destroyed by the heat of blanching, and blanching effectiveness is usually monitored by measuring peroxidase activity, in view of the greater heat resistance of this enzyme.

Williams et al.[106] recently considered some of the problems involved in using peroxidase as the universal indicator of blanching effectiveness. There is a substantial amount of evidence suggesting that the quality of products frozen after undergoing blanching is superior if a certain level of peroxidase activity remains at the end of the blanch.[107-110] Böttcher[109] concluded that the total absence of peroxidase activity was an indication of overblanching; for optimum product quality, he recommended blanching until the following levels of peroxidase activity: peas, 2–6·3%, according to the variety; green beans, 0·7–3·2%; cauliflower, 2·9–8·2%; and Brussels sprouts, 7·5–11·5%. Williams et al.[106] reported that peroxidase may not in fact be a good choice as indicator, inasmuch as its involvement in flavour and aroma deterioration has not been demonstrated. Furthermore, in most vegetables there are a number of isoenzymes of peroxidase with quite dissimilar heat stabilities (see Table 7), and complete inactivation requires considerably more heat treatment than that needed to inactivate other enzymes (see Table 8). Sensory analyses of English green peas and green beans have indicated that lipoxygenase is the key enzyme component in the development of undesirable odours, and all lipoxygenase activity ceases after about half the heating time required to bring about complete cessation of all peroxidase activity. A semi-quantitative method based on the conversion of

TABLE 7
TEMPERATURE STABILITY OF PEROXIDASE IN PLANT MATERIALS[a,b]

Plant material	z-value[c] (°F)	F-Value[d] (min at 82°C)
Asparagus	55–161	2–27
Beans, green	86–88	3·8
Beans, Lima	67	30
Corn[e]	70	—
Peas	48	60
Spinach[f]	59–81	6–10
Peaches	20–31	0·3–0·9
Pears	20–32	0·5–8·2
Tomato juice	18	0·9
Apples, fresh juice	11–34	5–7

[a] Adapted from Schwimmer.[111]
[b] McConnell,[112] unless otherwise stated.
[c] Slope of the logarithmic thermal inactivation curve.
[d] Time at 82°C to inactivate peroxidase (< 1% activity left).
[e] Vetter et al.[113]
[f] Duden and Fricker.[114]
Source: Williams et al.[106] Reproduced with permission of The Institute of Food Technologists.

TABLE 8

COMPARATIVE TEMPERATURE STABILITY OF SOME ENZYMES IN PLANT MATERIALS[a,b]

Enzyme	Food	z-value (average) (°F)	F-value (average) (min at 82°C)
Ascorbate oxidase	Peach and vegetables	59	2
Catalase	Vegetables[c]	28	6
Chlorophyllase	Spinach[d]	22	2
Lipoxygenase	Peas[e]	16	<0·1
Pectin esterase	Citrus juice	14	43
Peroxidase	Peas	48	60
Phosphatase	Orange juice	9	
	Milk	13	
Polygalacturonase	Citrus juice	16	12
	Papaya[e]	11	23
Polyphenol oxidase	Several fruits	12	1·1

[a] Adapted from Schwimmer.[111]
[b] McConnell,[112] unless otherwise stated.
[c] Sapers and Nickerson.[115]
[d] Resende et al.[36]
[e] Aylward and Haisman.[37]
Source: Williams et al.[106] Reproduced with permission of The Institute of Food Technologists.

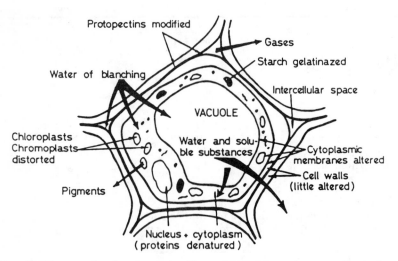

FIG. 7. Diagram showing the main effects of blanching on a generalized plant cell.[116]

iodide ions to iodine by the action of lipoxygenase on linoleic acid has been proposed as a measure of lipoxygenase activity and as an indicator of adequate blanching.

The heat produced during blanching kills the cells and solubilizes the pectic substances, causing irreversible alterations in cell structure and in the mechanical properties of plant tissues.[54] Figure 7 illustrates the main effects of blanching on a plant cell. Alterations increase the permeability of cytoplasmic membranes, allowing the blanch water to penetrate into the cells and intercellular spaces, driving out the gases and other volatile compounds. At the same time, proteins are denatured and soluble substances such as vitamins, mineral salts, and sugars, are lost. Chloroplasts and chromoplasts swell and rupture, the carotenes and chlorophylls diffusing into the cells and the blanching medium. Starches are similarly affected; they are solubilized and gelatinized, occupying all or part of the cell cytoplasm. These detrimental effects of blanching, mainly permanent alteration of plant tissue structure, solubilization and destruction of nutrients and vitamins in the blanching medium, and colour changes brought about by transformation of the chlorophylls into pheophytins, are more intense with longer blanching or higher blanching temperatures.[116]

In recent years the literature on blanching of vegetables has been considerable.[75-77,80,106,117-123] The work has tended to use different procedures to ameliorate at least partially the adverse effects of blanching, either by shortening blanching time or by palliating its detrimental action on sensory quality (texture, flavour, colour). Applying extremely short blanching times (thermal shock) of from 10–15 s using peas and green beans (Table 9), Steinbuch[124] obtained satisfactory colour and flavour results after a year of storage at −18°C, in spite of an unmistakable regeneration of polyphenol oxidase and even of catalase and lipoxygenase, and, in the case of green beans, texture was superior to that achieved with conventional blanching. These flavour findings were in disagreement with those obtained by Adams,[125] using blanching times of less than 30 s for peas and less than 1 min for green beans, in which a taste panel detected off-flavours after nine months in storage at −20°C. Dietrich and Newmann[34] proposed, for Brussels sprouts, whose multifoliate structure makes it difficult to achieve deep blanching without cooking the surface, a pre-heating treatment at 50°C in which the accumulated heat from the pre-heating stage did not damage tissues and allowed reductions of up to 20% in blanching time. Low-temperature (70°C) pre-treatment for extended periods (10–15 min) followed by cooling and brief, high-temperature (97°C) blanching reduced damage to the tissue

TABLE 9

RELATIONSHIP BETWEEN THE HEATING TIME, RESIDUAL ENZYME ACTIVITY, AND QUALITY RETENTION OF FROZEN PEAS AND FROZEN GREEN BEANS

| Heat treatment in water (98°C) | Residual enzyme activity (%) Peas | | | Quality evaluation after one year's storage | | | |
| | Lipoxygenase | Catalase | Peroxidase | Peas | | Green beans | |
				Colour	Flavour	Colour	Flavour	Texture
None	100	100	100	Discoloured	Strong off-flavour	Discoloured	Strong off-flavour	Good
2·5 s	80	36	65	Discoloured	Off-flavour	Discoloured	Off-flavour	Good
5 s	62	28	52	Discoloured	Good	Discoloured	Off-foavour	Good
10 s	6	2	34	Good	Good	Good	Good	Good
15 s	1	0·3	23	Good	Good	—	—	—
3 min	—	—	0·3	Good	Good	Good	Good	Soft

Source: Steinbuch[124]

structure. This step-wise blanching has yielded substantially improved final product texture in green beans,[126] potatoes,[54] carrots,[57] and peas,[59] although sizeable losses in soluble substances (vitamins and minerals) have been reported for carrots.[127] Utilization of mixed blanching methods using microwaving followed by immersion in boiling water reduced the duration of blanching in potatoes[128] and Brussels sprouts,[129] yielding products in which texture was more homogeneous and acceptable and vitamin C retention greater than in conventional blanching.

Lowering the pH by adding 0·5% citric acid to the blanch water increased the heat sensitivity of the enzyme systems, permitting reductions of between 20 and 30% in blanching time for artichokes.[95] Acidification of this kind is not, however, generally applicable, because it promotes the transformation of chlorophyll into pheophytin, thereby exerting an adverse effect on colour in green vegetables. On the other hand, addition of salts such as chlorides and sodium or potassium sulphate did not modify the pH while substantially decreasing the transformation of chlorophyll into pheophytin in spinach and Brussels sprouts. This beneficial effect was less pronounced in peas and green beans but continued through freezing and storage.[80,130] Adding calcium chloride or calcium citrates to the blanch water appreciably improved firmness in cauliflower and potatoes. The utilization of additives in the blanch water in order to improve vegetable quality or retain soluble substances should be the object of thorough study to determine all the possible beneficial and detrimental repercussions.[91] Appreciable losses in ascorbic acid and vitamins B_1, B_2, and niacin occur during blanching and the cooling that follows, particularly when these treatment steps are carried out in water.[131] Peas blanched for three minutes lose 33% of their initial ascorbic acid content, 20% of their riboflavin, 10% of their niacin, and 5% of their thiamine, even though these vitamins are stable during storage at temperatures of $-18°C$ or lower. Similarly losses of total sugars and soluble proteins can also be significant, depending upon the blanching method, duration, and temperature. Conventionally blanched peas suffer losses of 40% of their minerals, 30% of their sugars, and 20% of their proteins.[95] Blanching and the subsequent cooling are major sources of water pollution; and, despite the fact that these two operations use only 5–7 litres of water out of the total of 25–35 litres per kg of frozen vegetable, they are responsible for 60–70% of the total pollution.[118]

Considerable research and technological development took place in the field of energy use and environmental protection during the 1970s, in an attempt to improve the most frequently used water-based methods of blanching and cooling. Steam blanching at atmospheric pressure takes 30–50% longer than conventional water blanching yet produces 9–16 times fewer effluents. Blanching systems including water recycling,[132] steam blanching methods like fluidized bed blanching,[133] hydrostatically sealed steam blanching,[134] individual quick blanching (IQB),[135] and a spiral vibratory conveyor blanch–cooling system[136] all achieved faster, more uniform blanching with improved colour retention, considerably reduced nutrient loss, significantly lower leaching of solubles, energy savings, and better yields. Air cooling systems or systems using water sprays decreased the leaching of solubles and water pollution but resulted in high product weight loss, and for this reason they are not competitive with conventional cooling systems using running water.[118] Not all of the improved blanching methods developed have been implemented commercially. Poulsen[120] reviewed the status of vegetable blanching, comparing water and steam blanchers. He considered the cabinet integrated blancher–cooler equipped with a heat exchanger, developed by two Danish companies, Odense Cannery Ltd and Cabinplant International A/S, to be the best designed and most advanced solution on energy and pollution problems. This installation has a heating zone and a cooling zone. A pre-heating counterflow is produced in the heating zone, where the product is heated to 60°C, followed by blanching. Counterflow cooling of the product to 45°C followed by air-cooling then takes place in the cooling zone. Water heated in the cooling zone to about 80°C releases its heat to the water in the heating zone via a heat exchanger. This design achieves savings of about 60% of the calculated energy consumption and reduces water consumption by over 90%. The final air-cooling operation has the additional advantage of removing water before the product is conveyed to the freezer. Several such blanchers are now in operation in the UK, USA, New Zealand, Middle East, and Chile.[120]

Because of the detrimental effects of blanching on the product, energy costs and pollution, research has also been carried out into possible alternatives that can replace blanching without producing adverse alterations in product quality.[137–140] As a result of such research, onions, leeks, peppers, parsley, and cucumbers can be frozen without blanching with no appreciable loss in quality over relatively short storage periods (6–9 months). Except for these few products, blanching remains an essential step in the freezing process, and consequently research in this area should be continued in the future in order to optimize procedures and reduce the adverse effects on the final quality of frozen vegetable products.

4.2.1.3. Other protective treatments. Frozen fruits are normally consumed uncooked after gradual thawing, and blanching is not possible as a pre-freezing treatment. Certain berries, like blueberries and raspberries, can be frozen directly, without any protective pre-treatment. Because of their high acidity and high ascorbic acid content, the stability of these fruits during frozen storage is excellent. In contrast, apples, pears, and most stone fruits undergo quite rapid enzymatic browning and develop off-flavours after thawing. The methods used to prevent enzymatic browning by blocking the mechanisms involved in the oxidation–reduction of the polyphenols catalysed by polyphenol oxidase in the presence of oxygen and ascorbic acid include the addition of ascorbic acid, inhibition of polyphenol oxidase, or removal of free oxygen. Concentrations of from 0·1–0·3% ascorbic acid are sufficient, since enzymatic browning reactions are slow in an acid medium at low storage temperatures and reduced oxygen levels. Alternatively, peaches and apricots which have not been through an antioxygen pre-treatment can be treated during thawing by immersion in water or sugar syrup containing 0·1–0·5% ascorbic acid.[95] The chemicals most commonly used to inhibit the action of polyphenol oxidase include sulphur dioxide, citric acid, and sodium and calcium chloride. Frozen apples and apricots are commonly treated with sulphur dioxide by immersion in a 0·4–0·5% solution of SO_2 for 3–4 minutes. Tissues thus treated fix no more than 75 ppm (legal limit: 100 ppm). Concentrations of 1–3% citric acid provide temporary protection against enzymatic browning and microbial growth in peaches to be frozen. Brief immersion of *Champignon de Paris* minces in a solution of 0·8% citric acid and 1% NaCl significantly increased inhibition of enzymatic browning.[141] Saline solution concentrations ranging from 1–3% are applied previously to the antioxygen pre-treatment (immersion in a sugar syrup) to provide temporary protection against browning in sliced apples intended for freezing. The chloride anions, rather than the calcium cations, play the major role as inhibitors of browning.[142]

In a recent review of the literature, Espinosa[143] set forth the formulae of many of the solutions used for similar purposes in a variety of products, based on different concentrations of calcium chloride and sulphur dioxide suitable to each product at specified pH levels as combinations of chemical inhibitors applied prior to the final, definitive antioxygen treatment. Fruits are protected by the addition of sugar or immersion in a sugar syrup that acts as a protective agent thanks to its indirect antioxygen action of driving the oxygen out of cell tissues. Different concentrations have been recommended in a large number of papers, according to product type, generally running on the order of 30–40° Brix for cherries, pears, and plums, 40–50° Brix for bananas and apples, and 40–60° Brix for strawberries, peaches, and apricots.[95] As in blanching, adding calcium chloride to the sugar syrup also has a positive effect on texture in melons[144] and strawberries.[145] The addition of colloids (low methoxyl pectins, alginates, or agar extracts) in the proportion of 0·1–0·4% to the syrup improves product texture on thawing.

Research is required in order to optimize, in each individual case, the antioxygen treatments or the procedures intended to improve texture, given the wide variety of products and possible protective treatments.

4.2.2. Freezing Equipment
Holdsworth[146] recently published in *Developments in Food Preservation*, Volume 4, a full review of freezing methods and equipment along with other physical and engineering aspects of food freezing. The three most frequently used methods of freezing vegetables are direct contact (typically plate freezers), air-blast freezing systems, and cryogenic freezing. Plate freezers for vegetables are always horizontal and are used to freeze chopped or sliced products such as spinach in packages of up to 5 cm in height. The product is frozen by contact, with conduction as the main mode of heat transfer. The advantages of this method are economy and minimum weight loss, but freezing rates are moderately slow. Vegetables are preferably presented as individually quick-frozen products. IQF products are obtained in air-blast or cryogenic freezing systems. Air-blast freezing consists of blowing cold air through the product, which can be placed on trays in the case of a batch system (stationary, push-through, or automatic) or in continuous, moving belt systems. Belts may be arranged linearly, with a single or various belts one above the other, or in a rotting spiral, if the available floor space is limited. Normal operating conditions are between −30 and −40°C, with air velocities of up to 20 m/s. For proper operation, temperature and air velocity must be optimized as a function of product thickness. Precautions must be taken to reduce moisture loss (0·6–2%) and concomitant product dehydration and maintain surface quality.

In fluidized bed freezers, the cold air is used both for fluidization and freezing. Products entering the fluidized bed are frozen very quickly, with freezing times ranging from 3 min for peas to a maximum of 15 min for strawberries. Compared with moving belt freezers, fluidized bed freezing has the advantage of affording true individual quick freezing with lower weight loss, since each product particle is surrounded by a thin layer of frost that prevents dehydration. Freezing tunnels of this kind are suitable for foods with a uniform shape and diameters of up to 40mm and are most commonly used for vegetables such as peas, sliced green beans, sliced carrots, Brussels sprouts, corn, and Lima beans and for fruits like blueberries, sliced apple, and sliced pineapple, as well as for pre-fried potatoes. Throughputs of 2–5 tonnes/h can be achieved.

Tunnel freezers for products that are larger, cut into non-uniform shapes, or fragile, such as cauliflower, Brussels sprouts, and strawberries combine a first stage based on the fluidized bed principle, in which the product surface (crust freezing zone) is frozen with minimal weight loss, with a second stage in which products are deep frozen (completion freezing zone) on a conveyor belt.

The benefits of individual quick freezing in fluidized bed freezers as compared to conventional or moving belt freezing tunnels include greater heat transfer efficiency, and hence higher freezing rates with lower product weight loss and less frequent defrosting. Product handling is also facilitated, in that the product circulates freely through the tunnel, with considerable energy savings. The main drawback is that they cannot be used with large or non-uniform products.

More rapid freezing is achieved using cryogenic agents that boil at very low temperatures at atmospheric pressure—liquid nitrogen ($-196°C$) and LEF-R12 ($-30°C$)—in tunnels, cabinets, and spiral freezers. The cryogenic agent is lost to the atmosphere (LN_2) or recondensed (LEF-R12) after the vapours have been used to pre-cool and/or freeze the product. Cryogenic freezing results in rapid crust freezing and reduced weight loss (0·1–0·3%), impermeabilizing the product to oxygen, and causes minimal structural damage, yielding frozen products (particularly fruits) with excellent textural qualities. Cryogenic tunnels are relatively short and of simple construction and thus offer the advantages of low investment outlays and simple operation, making them economical for small, highly seasonal, high-value products despite the high cost of LN_2.[46] Recent cryogenic equipment designs have incorporated continuous monitoring, decreasing gas consumption and thereby lowering operating costs.[147] New designs have reduced the problems attaching to the different types of freezing equipment, e.g. products sticking to conveyor belts, and great attention has been given to hygiene, with food contact surfaces of stainless steel or high-quality plastic materials and incorporation of belt washing operations before loading with produce.[148]

The total freezing cost involves not only the capital expenditure and operating costs but also the cost of product weight losses. A classic comparison of the total freezing costs for the different systems indicated that the most economical were continuous air-blast systems as belt and fluidized bed freezers for freezing plants that enjoyed uniformly high production rates, as is the case for frozen vegetables. When freezing volume was not especially high, the operating costs of cryogenic systems dropped, along with their total costs, because of their low initial investment. Everington[149] concluded that, with the rapid growth of the value-added convenience food market, where the energy costs associated with freezing at air temperatures of $-30°C$ are only on the order of 0·2% of the sales price, the trend would be towards the use of lower refrigerant temperatures at the expense of energy costs. This trend is already observable in the growing use of cryogenic freezers, which have enjoyed considerable growth in the US freezing market.

4.3. Packaging

In additon to the physical and chemical changes causing the progressive deterioration of quality during storage, frozen fruits and vegetables can also suffer mechanical (breakage and disaggregation) and photochemical (colour and flavour denaturation) alterations. The extent of such alterations in quality is dependent in large measure upon the product preparation and the type of packaging employed. There are numerous papers dealing with the mechanical and physical properties of packaging materials but relatively few dealing with the effect of packaging type on the quality and stability of frozen vegetables.[150] The materials employed in packages for frozen foods should, first of all, possess all the usual features normally required of food packaging, namely, they should be chemically inert and stable, odour-free and odour-impervious, free of toxic substances that could be absorbed by food, impermeable to water vapour, volatile substances, and external odours, mechanically transformable into the appropriate size and shape for display in retail sales cabinets, easy to open, attractive, and capable of affording protection against microbial contamination.

In addition to these requirements, packaging materials for frozen foods should be shaped in such a way as to promote rapid freezing of the product inside, yet allow for expansion during freezing. They should also be impermeable to liquids, resistant to moisture, weak acids, and low temperatures, reflective and as opaque as possible, capable of conforming to product shape without adhering in order to prevent the air pockets that favour surface sublimation, and permeable and resistant to microwave energy in those cases in which reheating or cooking may be effected in microwave ovens.[46]

Frozen vegetables have their own special requirements for preparatory treatment and packaging. Certain products, such as cauliflower, Brussels sprouts, and cut green beans, are particularly fragile, calling for packages that can withstand the compression and shocks that occur during production. Ultraviolet radiation to a wavelength of 5000 Å can catalyse certain chemical reactions that can give rise to significant denaturation of colour in the case of chlorophyll-containing vegetables, making the use of opaque packaging materials essential.

Frozen vegetables undergo dehydration during storage, chiefly as a result of fluctuations in storage temperature and the degree of proofness of the packaging to water vapour. Such dehydration is irreversible, giving rise to ice formation inside the package and exerting detrimental effects on quality (alterations in colour and flavour, freezer burn, increased risk of oxidation, structural deterioration). Consequently, packages should ideally be air-tight, totally impermeable to water vapour, and effective as thermal insulators to limit possible temperature fluctuations within the product.

The alterations and losses in aroma and flavour, enzymatic browning, and oxidation of ascorbic acid that take place in the presence of oxygen call for the use of packaging materials that are air-tight (impervious to oxygen) or that permit removal of the oxygen from within the package by either creation of a partial vacuum or injection of inert gases (N_2 or CO_2).

Most vegetables are frozen using individual quick freezing (IQF) methods and stored in bulk containers for more or less protracted periods, after which they are repackaged in smaller retail packages. Various factors, primarily economic in nature, have led to the generalized use of polyethylene bags, even though they are ineffective in preventing mechanical damage and dehydration. The use of cardboard coated with paraffin or microcrystalline waves or plastics such as polyethylene or polypropylene or laminated films and foils impermeable to water vapour and oxygen (polyethylene- or polypropylene-coated Cellophane), aluminium foils laminated with plastic films, help prevent mechanical damage and dehydration. They also offer the additional advantage of being able to bear printing or contain a transparent window for viewing the product.

In response to the needs of catering services, the development of new plastic films during the 1960s in the US led to 'boil-in-the-bag' packages in which vegetables could be pre-cooked, frozen, and reheated. Such products are typically of high quality, and production is fully automated. Feinberg *et al.*[151] published a detailed description of this process, recently reviewed by Feldman *et al.*[152] The utilization of thermoplastic polyesters (Eastmant TENITE PET Polyester, 1987) with a very acceptable level of high-temperature dimensional stability and low-temperature toughness ($-80°C$ to $200°C$) has made possible 'freezer to oven to table' packaging in which products can be packaged prior to or after freezing, cooked in the package in conventional or microwave ovens on removal from the freezer, and served. For vegetables, this has permitted pre-cooked frozen products and new forms of preparation and packaging. There is a clear need for research to develop preparation procedures and packaging tailored to the many individual products.

4.4. Importance of PPP Factors
Earlier sections have highlighted the importance of PPP factors. As has already been pointed out, most frozen food legislation would appear to be excessively concerned with storage temperature while failing to give product, process, and packaging factors the importance they deserve. In

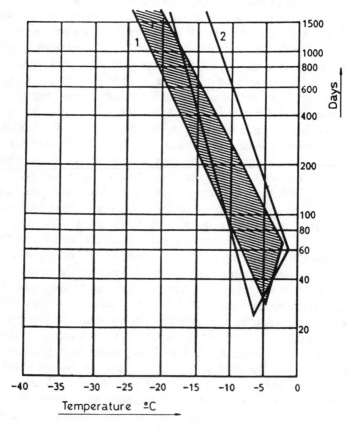

FIG. 8. Acceptability areas for vegetables and fruits and berries. (1) Vegetables, (2) fruits and berries. After Bengtsson et al.[153]

a recent review of the literature, Jul[62] found wide variability and great dissimilarity in the reported storage times, most likely ascribable primarily to initial differences in the PPP factors and the quality criteria applied. Bengtsson et al.[153] summarized the findings of various TTT experiments, and the acceptability areas for fruits, berries, and vegetables are illustrated in Fig. 8. From the Albany TTT Project to the COST 91 Project, relatively few TTT/PPP experiments have been conducted on frozen vegetables. Bøgh–Sørensen[154] summarized the reports of such experiments in the framework of the COST 91 Project dealing with the effect of reducing or omitting blanching,[155,156] optimum harvesting times for different varieties,[157] and sensory and objective measurements of colour,[158] all on green beans.

The effects of stepwise blanching on potatoes, carrots, and peas,[59] blanching of parsley,[159,160] and several pre-treatments on mushrooms, sliced apples, and strawberries[161] have also been studied. The use of vacuum packaging on blanched and unblanched carrots was studied by Espinosa et al.,[162] on blanched cauliflower by Fuster et al.[163]. All of these studies recommended improving the initial product quality, processing, or packaging in order to achieve longer storage lives, and they attest to the need for more research aimed at determining the influence of each of the individual PPP factors and the optimum combinations thereof, with a view to achieving more representative maximum acceptability and stability times for the various products or product groups.

4.5. Microbiological and Nutritional Aspects

Microorganisms are inevitably present on raw vegetable materials. Freezing and cold storage kills some of these microorganisms, but this is a very slow and variable process, dependent upon product type and the type of microbial contamination. Since freezing is unable to reduce microbial contamination to any significant extent, the initial sanitary condition of the product is of the utmost importance. Table 10 shows the effect of the various operations in the freezing process on the initial bacterial count in peas. The count can be seen to drop sharply with washing and blanching, only to rise again, which points out the need to reduce all possible risk of recontamination between blanching and freezing. Special care must be taken when washing leafy products like spinach. Microorganism growth does not take place at the very cold temperatures employed in frozen storage (-18 to $-30°C$). During transport, distribution, and sale, storage temperatures should be kept below $-10°C$, below which no bacterial growth occurs, and no part of the product should be allowed to warm to a temperature above $-12°C$.[46]

TABLE 10

EFFECT OF PROCESSING ON BACTERIAL COUNT OF PEAS AT VARIOUS
STAGES OF PROCESSING

Point of sampling	Thousands of bacteria per gram of peas
Platform	11,346
After washing	1,090
After blanching	10
End of flume	239
End of inspection belt	410
Entrance to freezer	736
After freezing	560

Source: Peterson and Gunderson.[164]

Turning to nutrition, care should be taken, as was already pointed out in preceding sections, to prevent vitamin and other nutrient losses, basically between harvesting and processing, during blanching, and during frozen storage. Losses sustained during the time prior to processing depend upon the amount of time elapsing and the temperature. Losses occurring during blanching amount to between 10 and 30% of vitamins B and C, depending upon the blanching medium, blanching time, and preparatory processing undergone by the product. The nutritive value of the macronutrients does not practically decrease during storage at temperatures below $-18°C$ for periods of up to one year. Vitamin C remains rather stable over a storage period of one year if product temperature is colder than $-18°C$, with losses associated with temperature and varying with product type ($Q_{10} = 6$–20 in vegetables and $Q_{10} = 30$–70 in fruits).[165] Vitamins of the B-complex are more stable than vitamin C, except for folic acid, which is similar suffering moderate losses (20% over one year) at $-18°C$. Losses of carotene are minor (5–20%). The final cooking of frozen vegetables can produce substantial losses: 10–50% of the vitamin C, 0–50% of the folic acid, 0–25% of the vitamin B_1, 20–40% of the vitamin B_6, and 10–40% of the pantothenic acid.[165]

Finally, available experimental data on table-ready dishes prepared from garden peas indicated that vegetable freezing was less destructive from the nutritional standpoint (vitamin C loss $=61\%$) than other processing methods such as canning (64%), air-drying (75%), or freeze-drying (65%). The frozen product also compared favourably with the raw, chilled equivalent, purchased at the market, in which the vitamin C loss was 56%.[166,167]

5. OUTLOOK FOR THE FUTURE

This summary of research in the field of frozen vegetable products and product quality has highlighted the diversity of the factors involved. Research has generally concentrated upon the changes undergone by products under the influence of variations in different stages of the freezing process or in storage conditions, but the 'how and why' of such changes has often gone unexplained. Modifications to processes should be based on basic research into the physical, chemical, and biological features of phytosystems at low temperatures to improve our understanding of the behavioural processes taking place during freezing. A review of the recent literature indicates the need for research in the following areas:

(1) The selection and breeding of high-yield varieties possessing appropriate quality attributes for freezing and for mechanical harvesting and processing are of great importance. The development of objective methods for measuring product ripeness, in order to allow proper quality control prior to processing, is another important task.

(2) Determination of the physical properties of the various products, in particular their mechanical and thermophysical properties, before, during, and after freezing, since these exert a major influence on process optimization and control and processing equipment design.

(3) The influence of the thermal treatments, including blanching, freezing, thawing, and/or cooking, on quality attributes, particularly to develop objective methods of measuring and analysing chemical and biochemical changes in structural components (in the case of texture) and in key compounds (in the case of alterations in colour and development of off-flavours and -odours).

(4) Freezing and its effects on plant tissues, in order to understand better the mechanical action of ice crystals on structure, the physico-chemical changes taking place and kinetics of enzymatic and non-enzymatic reactions at low temperatures.

(5) The extent to which ultra-fast freezing rates are beneficial to product quality, how long such beneficial effects last during storage, and to what degree they are detectable after cooking.

(6) Enzyme systems involved in changes in colour, flavour, and aroma, to establish optimum conditions for protective thermal or chemical treatments to yield the desired level of product quality at the end of storage.

(7) The effect of different thermal inactivation methods and possible alternative chemical means, to replace blanching or reduce its detrimental effects on quality.

The variability obtained up to now in TTT studies, due primarily to differences in PPP factors and the quality analysis methods employed, calls for, first, further investigation into stringent product, process, and packaging standards, and, second, standardization of objective methods of quality assessment and development of more precise and representative methods of sensory evaluation. The possibility of using higher storage temperatures holds out great promise for the future with a view to energy conservation. To this end, TTT/PPP studies are needed in order to classify products according to their stability during storage at different temperatures. Such studies should also consider the influence of temperatures under actual freezer chain conditions, taking into account the effect of temperature fluctuations on quality. The aim of this of necessity incomplete list of future research tasks is the ongoing improvement of product quality. The development of new generations of frozen fruit and vegetable products with higher values added and competitive pricing will depend upon the creative efforts and technological development stemming from cooperation between scientists and manufacturers, as they strive to increase the competitiveness of this important sector of the food industry and thereby to satisfy ever higher consumer expectations.

REFERENCES

1. MARIANI, A. In *Thermal Processing and Quality of Foods* (ed. by P. Zeuthen, J. C. Cheftel, C. Eriksson, M. Jul, H. Leniger, P. Linko and G. Varela), Elsevier Applied Science, London, 1984, p. 819.
2. ELMAN, F. *La Surgelation*, 1985, **240**, 27.
3. HALIQUE, V. *La Surgelation*, 1987, **263**, 11.
4. OLSON, R. L. In *Low Temperature Biology of Foodstuffs* (ed. by J. Hawthorn and E. J. Rolfe), Pergamon Press, London, 1968, p. 381.
5. TRESSLER, D. K. and EVERS, C. F. *The Freezing Preservation of Foods*, Avi Publishing Co., New York, 1957.
6. JOSLYN, M. A. and CRUESS, W. V. *Fruit Products J.*, 1929, **8**, 9.
7. ONSLOW, M. W. *Principles of Plant-Biochemistry*, Cambridge University Press, 1931.
8. JOSLYN, M. A. *Ind. and Eng. Chem.*, 1941, **33**, 308.
9. WOODROOF, J. G. *Ga. Inst. Technol., Eng., Expl. Sta. Bull.*, 1938, No. 201.
10. MORRIS, T. N. and BARKER, J. *Rep. Fd Invest. Board, (UK)*, 1942, **92**, 4.
11. LEE, F. A. and GORTNER, W. A. *Refrig. Eng.*, 1949, **57**, 148.
12. LEE, F. A., GORTNER, W. A. and WHITCOMBE, J. *Food Technol.*, 1949, 3(5), 164.
13. ARSDEL, W. B. VAN, COPLEY, M. J. and OLSON, R. L. In *Quality and Stability of Frozen Foods* (ed. by W. B. van Arsdel, M. J. Copley and R. L. Olson), Wiley Interscience, New York, 1969.

14. OLSON, R. L. and DIETRICH, W. C. In *Quality and Stability of Frozen Foods* (ed. by W. B. van Arsdel, M. J. Copley and R. L. Olson), Wiley Interscience, New York, 1969, pp. 117, 124.
15. GUADAGNI, D. G. In *Quality and Stability of Frozen Foods* (ed. by W. B. van Arsdel, M. J. Copley and R. L. Olson), Wiley Interscience, New York, 1969, p. 85.
16. GUADAGNI, D. C. In *Low Temperature Biology of Foodstuffs* (ed. by J. Hawthorn and E. J. Rolfe), Pergamon Press, London, 1968, p. 399.
17. WEIER, T. E. and STOCKING, C. R. *Adv. Food Res.*, 1949, **2**, 297.
18. BASSI, M. and CRIVELLI, M. G. *Rev. Gén. Froid*, 1968, **59**, 1211.
19. BASSI, M. and CRIVELLI, M. G. *Rev. Gén. Froid*, 1969, **60**, 1289.
20. CRIVELLI, M. G. and BASSI, M. *Rev. Gén. Froid*, 1969, **60**, 241.
21. REEVE, R. M. *J. Text. Stud.*, 1970, **1**, 247.
22. BROWN, M. S. *J. Sci, Food Agric.*, 1967, **18**, 77.
23. RAHMAN, A. R., HENNING, W. L. and WESTCOTT, D. E. *J. Food Sci.*, 1971, **36**, 500.
24. MOHR, W. P. *J. Text. Stud.*, 1974, **5**, 13.
25. JOSLYN, M. A. *Adv. Enzymol.*, 1949, **9**, 613.
26. LEE, F. A., WAGENKNECHT, A. C. and HENNING, J. C. *Food Res.*, 1956, **20**, 289.
27. LEESON, J. A. *Sci. Bull.*, Campden Food Preservation Research Association, Chipping Campden, Glos., UK, 1957, **2**.
28. LEE, F. A. *Adv. Food Res.*, 1958, **8**, 63.
29. ESSLEN, W. B. and ANDERSON, E. E. *Food Res.*, 1956, **21**, 322.
30. ZOUEIL, M. E. and ESSLEN, W. B. *Food Res.*, 1959, **24**, 119.
31. SAPERS, G. M. and NICKERSON, J. T. R. *J. Food Sci.*, 1962, **27**, 287.
32. DIETRICH, W. C., LINDQUIST, F. E., BOHART, G. S. and MORRIS, H. J. *Food Res.*, 1955, **20**, 480.
33. PINSENT, B. R. W. *J. Food Sci.*, 1962, **27**, 120.
34. DIETRICH, W. C. and NEWMANN, H. J. *Food Technol.*, 1965, **19**, 1174.
35. ROSOFF, H. D. and CRUESS, W. V. *Food Res.*, 1949, **14**, 283.
36. RESENDE, R., FRANCIS, F. J. and STUMBO, C. R. *Food Technol.*, 1969, **23**(1), 63.
37. AYLWARD, F. and HAISMAN, D. R. *Adv. Food Res.*, 1969, **17**, 1.
38. DALHOFF, E. and JUL, M. *Progress in Refrigeration Science and Technology*, Pergamon Press, New York, 1965, **1**, 57.
39. GUTSCHMIDT, J. In *Low Temperature Biology of Foodstuffs* (ed. by J. Hawthorn and E. J. Rolfe), Pergamon Press, London, 1968, p. 299.
40. OLSON, R. L. and DIETRICH, W. C. In *The Freezing Preservation of Foods* (ed. by D. K. Tressler, W. B. van Arsdel and M. J. Copley), 4th edn, Avi Publishing Co., Inc., Westport, Connecticut, 1968, p. 83.
41. POINTINE, J. D., FEINBERG, B. and BOYLE, F. P. In *The Freezing Preservation of Foods* (ed. by D. K. Tressler, W. B. van Arsdel and M. J. Copley), 4th edn, Avi Publishing Co., Inc., Westport, Connecticut, 1968, p. 107.
42. FENNEMA, O. R., POWRIE, W. D. and MARTH, E. H. In *Low Temperature Preservation of Foods and Living Matter* (ed. by O. R. Fennema), Marcel Dekker, Inc., New York, 1973, pp. 202, 207, 380.
43. CIOUBANU, A., LASCU, G., BERCESCU, V. and NICULESCU, L. *Cooling Technology in the Food Industry*, Abacus Press, Kent, UK, 1976.

44. DESROSIER, N. W. and TRESSLER, D. K. In *Fundamentals of Food Freezing* (ed. by N. W. Desrosier and D. K. Tressler), Avi Publishing Co., Inc., Westport, Connecticut, 1977.

45. INTERNATIONAL INSTITUTE OF REFRIGERATION, *Recommendations for the Processing and Handling of Frozen Foods* (ed. by International Institute of Refrigeration), 2nd edn, 1972, pp. 82, 36.

46. INTERNATIONAL INSTITUTE OF REFRIGERATION, *Recommendations for the Processing and Handling of Frozen Foods* (ed. by International Institute of Refrigeration), 3rd edn, 1986, pp. 32, 34, 38, 236, 240, 264, 266.

47. ILDE, D. B. and HUDSON, M. A. In *Low Temperature Biology of Foodstuffs* (ed. by J. Hawthorn and E. J. Rolfe), Pergamon Press, London, 1968, p. 153.

48. FENNEMA, O. R. In *Principles of Food Science*, Part II (ed. by O. R. Fennema), Marcel Dekker, Inc., New York, 1976, p. 173.

49. GUTSCHMIDT, J. In *Low Temperature Biology of Foodstuffs* (ed. by J. Hawthorn and E. J. Rolfe), Pergamon Press, London, 1968, p. 299.

50. BROWN, D. C. *Advan. Cryogenic Eng.*, 1967, **12**, 11.

51. REEVE, R. M. *J. Agric. Food Chem.*, 1972, **20**, 1982.

52. BROWN, M. J. *J. Text. Stud.*, 1977, **7**, 391.

53. BROWN, M. J. *Adv. Food Res.*, 1979, **25**, 181.

54. CANET, W. Estudio de la influencia de los tratamientos térmicos de escaldado, congelación y descongelación en la textura y estructura de patata, Ph.D. thesis, 1980, Universidad de Madrid.

55. CANET, W., ESPINOSA, J. and RUIZ ALTISENT, M. In *Progress in the Design and Operation of Refrigerating Equipment and in the Processing of Fruits and Vegetables by Refrigeration*, Refrigeration Science and Technology, Sofia, Bulgaria, 1982, p. 277.

56. CANET, W., ESPINOSA, J. and RUIZ ALTISENT, M. In *Progress in the Design and Operation of Refrigerating Equipment and in the Processing of Fruits and Vegetables by Refrigeration*, Refrigeration Science and Technology, Sofia, Bulgaria, 1982, p. 284.

57. CANET, W. and ESPINOSA, J. *Rev. Agroquim. Technol. Aliment.*, 1983 **23**(4), 531.

58. CANET, W. and ESPINOSA, J. *Actas Ier Congresso Nacional de Ciencias Horticolas*, Sociedad Española de Ciencias Horticolas, Valencia, 1983, **II**, 943.

59. CANET, W. and ESPINOSA, J. In *Thermal Processing and Quality of Foods* (ed. by P. Zeuthen, J. C. Cheftel, C. Eriksson, M. Jul, H. Leniger, P. Linko and G. Varela), Elsevier Applied Science, London, 1984, p. 678.

60. MONZINI, A., CRIVELLI, G., BASSI, M. and BUONOCORE, C. *Structure of Vegetables and Modifications due to Freezing*, Instituto Sperimentale per la Valorizzazione Tecnologica dei Prodotti Agricoli, Milan, 1975, p. 52.

61. BROWN, M. J. *Proc. XIII Int. Congr. Refrig.*, 1971, **3**, 491.

62. JUL, M. In *The Quality of Frozen Foods* (ed. by M. Jul), Academic Press, Inc., London, 1984, pp. 20, 53, 58, 242, 243, 208.

63. DELAUNAY, J. and ROSSET, R. *Deep Frozen Products*, Informations des Services Vétérinaires, Paris, 1981, Nos, 72–75, 21.

64. ULRICH, R. *Rev. Gén. Froid*, 1981, **71**(7/8), 371.

65. VEERKAMP, C. H. In *Thermal Processing and Quality of Foods* (ed. by P. Zeuthen, J. C. Cheftel, C. Eriksson, M. Jul, H. Leniger, P. Linko and G. Varela), Elsevier Applied Science, London, 1984, p. 802.

66. CALVELO, A. In *Developments in Meat Science* (ed. by R. Lawrie), Applied Science Publishers Ltd, London, 1981, **2**, p. 125.
67. ÅSTRÖM, St. *Proceedings Symposium on Frozen and Quick Frozen Food — New Aspects for Agricultural Production and Marketing*, FAO, New York, 1977, p. 149.
68. ZARITZKY, N. E., ANN, M. C. and CALVELO, A. *Meat Sci.*, 1982, **7**, 299.
69. CANET, W. *Alim. Equip. Tech.*, 1986, **2**, 125.
70. JOSLYN, M. A. In *Cryobiology* (ed. by H. T. Meryman), Academic Press, New York, 1966, p. 565.
71. LEE, C. Y., BOURNE, M. C. and VAN BUREN, J. P. *J. Food Sci.*, 1979, **44**(2), 615.
72. BENGTSSON, B. L. and BOSUND, I. *J. Food Sci.*, 1966, **31**, 474.
73. RHEE, K. S. and WATTS, B. M. *J. Food Sci.*, 1966, **31**(5), 675.
74. MURRAY, K. E., SHIPTON, J., WHITFIELD, F. B. and LAST, J. H. *J. Sci. Food Agric.*, 1976, **27**, 1093.
75. ULRICH, R. *Rev. Gén. Froid*, 1983, **73**, 11.
76. PHILIPPON, J. and ROUET-MAYER, M. A. *Int. J. Refrig.*, 1985, **8**(2), 254.
77. CANET, W. *Alim. Equip. Tech.*, 1986, **5**, 77.
78. BUCKLE, K. A. and EDWARDS, R. A. *J. Sci. Food Agric.*, 1970, **21**, 307.
79. WALKER, G. C. *J. Food Sci.*, 1964, **29**, 389.
80. PHILIPPON, J. and ROUET-MAYER, M. A. *Int. J. Refrig.*, 1985, **8**(1), 48.
81. BOGGS, M. M. and DIETRICH, W. C. *Food Technol.*, 1960, **14**, 181.
82. KRINSKI, N. I. In *Biochemistry of Chloroplasts* (ed. by I. W. Goodnin), Academic Press, New York, 1966, p. 423.
83. JASWAL, A. S. *Amer. Potato J.*, 1970, **47**, 145.
84. DIETRICH, W. C., BOGGS, M. M., NUTTING, M. D. and WEINSTEIN, N. E. *Food Technol.*, 1960, **14**, 522.
85. POLENSKY, W. *Industr. Obst. Gemüsevrert.*, 1979, **64**, 577.
86. BERG, L. VAN DEN *Food Sci.*, 1964, **29**, 540.
87. BERG, L. VAN DEN *Food Technol.*, 1961, **15**, 434.
88. ARSDEL, W. B. VAN *Fd. Technol.* 1957, **11**(1), 28.
89. DIETRICH, W. C., NUTTING, M. D., BOGGS, M. M. and WEINSTEIN, N. E. *Food Technol.*, 1962, **16**(10), 123.
90. SPIESS, W. E. L. In *The Quality of Frozen Foods* (ed. by M. Jul), Academic Press, Inc., London, 1984, p. 242.
91. MUÑOZ-DELGADO, J. A. *Refrigeración y Congelación de Alimentos Vegetales*, Fundación Española de la Nutrición, Serie Informes No. 2, 1985, p. 83.
92. STEINBUCH, E., SPIESS, W. E. L. and GRÜNEWALD, Th. *Bull. Int. Inst. Refrig.*, Annex, 1977, **1**, 239.
93. BERG, L. VAN DEN *Food in Canada*, 1966, **26**(5), 37.
94. HAWKINS, A. E., PEARSON, C. A. and RAYNOR, D. *Proc. Inst. Refrig. (UK)*, 1973, 69.
95. PHILIPPON, J. *Bull. Tech. Ind. M6–AGRO–64*, 1975, **296**, 81.
96. PHILIPPON, J. *Rev. Prat. Froid*, 1967, **20**, 261.
97. HERRMANN, K. *Ernährungsumschau*, 1970, **17**, 458.
98. CRIVELLI, G., FIDEGHELLI, C. and MONASTRA, F. *Proc. XIV Int. Congre. Refrig.*, 1975, **3**, 546.
99. CRIVELLI, G. and SCOZZOLI, R. *Proc. XIV Int. Cong. Refrig.*, 1975, **3**, 550.
100. COUSIN, R. *Compt. Rend. Colloque CENECA, le Froid en Agriculture*, Rapport No. 5225, Paris, 1976.

48   W. CANET

101. LATRASSE, A. *Compt. Rend. Colloque CENECA, le Froid en Agriculture,* Rapport No. 5228, Paris, 1976.
102. RISSER, G., VAILLEN, J., FERRY, P. and CABIBER, M. *Compt. Rend. Colloque CENECA, le Froid en Agriculture,* Rapport No. 5227, Paris, 1976.
103. MUÑOZ-DELGADO, J. A. In *Food Quality and Nutrition* (ed. by W. K. Downey), Applied Science Publishers Ltd, London, 1977, p. 353.
104. CIOUBANU, A. and NICULESCU, L. In *Cooling Technology in the Food Industry* (ed. by A. Cioubanu, G. Lascu, V. Bercescu and L. Niculescu), Abacus Press, Kent, UK, 1976, p. 377.
105. GRAHAM, R. P. *Food Technol.,* 1969, **23**, 61.
106. WILLIAMS, D. C., LIM, M. H., CHEN, A. O., PANGBORN, R. M. and WHITAKER, J. R. *Food Technol.,* 1986 **40**(6), 130.
107. CAMPBELL, H. *West. Canner Packer,* 1940, **32**(9), 51.
108. WINTER, E. Z. *Lebensm. Unters. Forsch.,* 1969, **141**, 201.
109. BÖTTCHER, H. *Nahrung,* 1975, **19**, 173.
110. DELINCÉE, H. and SCHAEFER, W. *Lebensm. Wiss. Technol.,* 1975, **8**, 217.
111. SCHWIMMER, S. *Source Book of Food Enzymology,* Avi Publishing Co., Inc., Westport, Connecticut, 1981, p. 202.
112. MCCONNELL, J. E. W. *Natl. Canners Assoc., Publ. D-252,* 1956, May 25.
113. VETTER, J. L., NELSON, A. I. and STEINBERG, M. P. *Food Technol.,* 1959, **13**, 410.
114. DUDEN, R. and FRICKER, H. *Lebensm. Wiss. Technol.,* 1975, **8**, 147.
115. SAPERS, G. M. and NICKERSON, J. T. R. *J. Food Sci.,* 1962, **27**, 277.
116. KATSABOXAKIS, K. Z. In *Thermal Processing and Quality of Foods* (ed. by P. Zeuthen, J. C. Cheftel, C. Eriksson, M. Jul, H. Leniger, P. Linko and G. Varela), Elsevier Applied Science, London, 1984, p. 559.
117. P. Zeuthen, J. C. Cheftel, C. Eriksson, M. Jul, H. Leniger, P. Linko and G. Varela (eds), *Thermal Processing and Quality of Foods,* Elsevier Applied Science, London, 1984, p. 499.
118. PHILIPPON, J. *Sci. Aliments.,* 1984, **4**, 523.
119. PHILIPPON, J. and ROUET-MAYER, M. A. *Int. J. Refrig.,* 1984, **7**(6), 384.
120. POULSEN, K. P. *Food Technol.,* 1986, **40**(6), 122.
121. PIZZOCARO, F., GASPAROLI, A., BOSCHETTI, T. and MONTEVERDI, R. *Proc. XVII Int. Congr. Refrig.,* Commission C2, 1987, p. 447.
122. SELMAN, J. D. In *Developments in Food Preservation* (ed. by S. Thorne), Elsevier Applied Science, London, 1987, Vol. 4, p. 205.
123. SENESI, E., BERTOLO, G. and MAESTRELLI, A. *Proc. XVII Int. Congr. Refrig.,* Commission C2, 1987, p. 455.
124. STEINBUCH, E. In *Thermal Processing and Quality of Foods* (ed. by P. Zeuthen, J. C. Cheftel, C. Eriksson, M. Jul, H. Leniger, P. Linko and G. Varela), Elsevier Applied Science, London, 1984, p. 553.
125. ADAMS, J. B. *Rev. Gén. Froid,* 1983, **1**, 21.
126. STEINBUCH, E. *J. Food Technol.,* 1977, **11**, 313.
127. PALA, M. *Proc. XVI Int. Congr. Refrig.,* 1983, **3**, 631.
128. CANET, W. and HILL, M. *Int. J. Food Sci. Tech.,* 1987, **22**, 273.
129. CANET, W., GIL, M. and ALIQUE, R. *Int. J. Food Sci. Tech.,* submitted.
130. HUDSON, M. A., SHARPLESS, V. J., PICKFORD, E. and LEACH, N. *J. Food Technol.,* 1974, **9**, 95–103, 105–114.

131. PAULUS, K., DUDEN, R., FRICKER, A., HEINTZO, K. and ZOHM, H. *Lebensm. Wiss. Technol.*, 1975, **8**, 11.
132. SWARTS, J. G. and CORROAD, P. A. *J. Food Sci.*, 1981, **46**, 440.
133. MITCHELL, R. S., BOARD, P. W. and LYNCH, L. S. *Food Technol.*, 1968, **22**, 717.
134. LEE, C. Y. *Korean J. Food Sci. Technol.*, 1975, **7**(2), 100.
135. LAZAR, M. E., LUND, B. and DIETRICH, W. C. *Food Technol.*, 1971, **25**, 684.
136. BROWN, G. E., BOWBEN, J. L., DIETRICH, W. C., HUDSON, J. S. and FARKAS, D. F. *J. Food Sci.*, 1974, **39**, 696.
137. SULC, S. *Proc. XIV Int. Congr. Refrig.*, 1975, **3**, 525.
138. KOZLOWSKI, A. V. *Bull. Int. Inst. Refrig.*, Annex, 1977, **1**, 227.
139. STEINBUCH, E. *J. Food Technol.*, 1979, **14**, 321.
140. MUFTUGIL, N. and YIGIT, V. *Proc. XVI Int. Cong. Refrig.*, 1983, **3**, 613.
141. ROUET-MAYER, M. A., PHILIPPON, J. and DERENS, E. *Proc. XVII Int. Congr. Refrig.*, Commission C2, 1987, p. 279.
142. PHILIPPON, J., ROUET-MAYER, M. A., and BROGHIER, J. J. *Bull. Int. Inst. Froid.*, Annex, 1978, **2**, 29.
143. ESPINOSA, J. In *Thermal Processing and Quality of Foods* (ed. by P. Zeuthen, J. C. Cheftel, C. Eriksson, M. Jul, H. Leniger, P. Linko and G. Varela), Elsevier Applied Science, London, 1984, p. 533.
144. RIO, M. A. and MILLER, M. W. *Proc. XV Int. Congr. Refrig.*, 1979, **3**, 923.
145. WEGENER, J. B., BAER, B. H. and RODGERS, F. D. *Food Technol.*, 1951, **5**, 76.
146. HOLDSWORTH, S. D. In *Developments in Food Preservation* (ed. by S. Thorne), Elsevier Applied Science, London, 1987, Vol. 4, p. 153.
147. WILLHOFT, E. M. A. *Contributions to the Symposium on Preparation, Processing and Freezing in Frozen Food Production*, Food Engineering Forum, Institute of Mechanical Engineers, London, October, 1986.
148. CAMPBELL-PLATT, G. In *Food Technology International Europe 1987* (ed. by A. Turner), 1987, p. 63.
149. EVERINGTON, D W. *Contributions to the Symposium on Preparation, Processing and Freezing in Frozen Food Production*, Food Engineering Forum, Institute of Mechanical Engineers, London, October, 1986.
150. PHILIPPON, J. *Rev. Gén. Froid*, 1981, **3**, 1968, 127.
151. FEINBERG, B., WINTER, F. and ROTH, T. L. In *The Freezing Preservation of Foods*, (ed. by D. K. Tressler, W. B. van Arsdel and M. J. Copley) (4th edn), Avi Publishing Co., Inc., Westport, Connecticut, **3**, p. 150.
152. FELDMAN, C., ROSSET, R. and POUMETROL, C. *Rev. Gén. Froid*, 1983, **2**, 109.
153. BENGTSSON, N., LILJEMARK, A., OLSSON, P. and NILSSON, B. *Bull. Int. Inst. Refrig.* Annex, 1972, **2**, 303.
154. BØGH-SØRENSEN, L. In *Thermal Processing and Quality of Foods* (ed. by P. Zeuthen, J. C. Cheftel, C. Eriksson, M. Jul, H. Leniger, P. Linko and G. Varela), Elsevier Applied Science, London, 1984, p. 511.
155. KATSABOXAKIS, K. Z. and PAPANICOLAU, D. N. In *Thermal Processing and Quality of Foods* (ed. by P. Zeuthen, J. C. Cheftel, C. Eriksson, M. Jul. H. Leniger, P. Linko and G. Varela), Elsevier Applied Science, London, 1984, p. 684.
156. PHILIPPON, J., ROUET-MAYER, M. A. and ABBAS, J. In *Thermal Processing and Quality of Foods* (ed. by P. Zeuthen, J. C. Cheftel, C. Eriksson, M. Jul, H.

Leniger, P. Linko and G. Varela), Elsevier Applied Science, London, 1984, p. 729.

157. CRIVELLI, G., BERTOLO, G., MAESTRELLI, A. and SENESI, E. In *Thermal Processing and Quality of Foods* (ed. by P. Zeuthen, J. C. Cheftel, C. Eriksson, M. Jul, H. Leniger, P. Linko and G. Varela), Elsevier Applied Science, London, 1984, p. 696.

158. ADAMS, J. B. and ROBERTSON, A. In *Thermal Processing and Quality of Foods* (ed. by P. Zeuthen, J. C. Cheftel, C. Eriksson, M. Jul, H. Leniger, P. Linko and G. Varela), Elsevier Applied Science, London, 1984, p. 792.

159. PHILIPPON, J., ROUET-MAYER, M. A., FONTENAY, P. and DUMINIL, J. *Sci. Aliments*, 1986, **6**, 433.

160. ROUET-MAYER, M. A., PHILIPPON, J., DUMINIL, J. M. and FONTENAY, P. *Sci. Aliments*, 1986, **6**, 233.

161. FUSTER, C., PRESTAMO, G. and ESPINOSA, J. In *Thermal Processing and Quality of Foods* (ed. by P. Zeuthen, J. C. Cheftel, C. Eriksson, M. Jul, H. Leniger, P. Linko and G. Varela), Elsevier Applied Science, London, 1984, p. 671.

162. ESPINOSA, J., PRESTAMO, G., FUSTER, C. and CANET, W. *Proc. XVI Int. Congr. Refrig.*, 1983, **3**, 611.

163 FUSTER, C., PRESTAMO, G., CANET, W. and ESPINOSA, J. *Proc. XVI Int. Congr. Refrig.*, 1983, **3**, 575.

164. PETERSON, A. C. and GUNDERSON, M. F. In *The Freezing Preservation of Foods* (ed. by D. K. Tressler, W. B. van Arsdel and M. J. Copley), Avi Publishing Co., Inc., Westport, Connecticut, 1968, **2**, p. 289.

165. HARRIS, R. S. and KARMAS, E. *Nutritional Evolution of Food Processing*, Avi Publishing Co., Inc., Westport, Connecticut, 1975, p. 70.

166. BENDER, A. E. In *Food Technology International Europe 1987* (ed. by A. Turner), 1987, p. 273.

167. BENDER, A. E. In *Developments in Food Preservation* (ed. by S. Thorne), Elsevier Applied Science, London, 1987, Vol. 4, p. 1.

Chapter 2

FLAME STERILISATION OF CANNED FOODS

J. R. HEIL

Department of Food Science and Technology, University of California, USA

SUMMARY

Flame sterilisation is a precision, high temperature-short time process in the container. Suitable foods processed by flame sterilisation retain more quality, freshness, and nutritive value than foods thermally processed in cans by other methods. The flame process is adaptable to two packing styles: conventional syrup/brine packs and vacuum packs. Flame sterilised vacuum packed foods, with potential for heat preserved natural foods without salt or sugar added, are currently being evaluated for commercial use.

Process monitoring systems developed for flame sterilisation are capable of measuring adequacy of process delivered to each can. Cost of the flame steriliser is competitive with the cost of any comparable conventional processing unit. Energy usage of the flame steriliser, as determined by several research groups, varies with sources of information. Heat transfer in flame and potential improvements of flame steriliser efficiency are discussed.

NOTATION

A_c	External area of can (m^2)
A_{clm}	Logarithmic mean cylindrical area of can (m^2)
A_f	Area of flame (m^2)
A_i	Inside area of can (m^2)
a	Initial number of concentration heat labile factor before heating
α_c	Absorptivity of can, not dimensional (n.d.)
b	Number or concentration of heat labile factor after heat treatment

C Specific heat of food (J/kg K)

C_{pg} Average specific heat of combustion gases (J/kg mol K)

D Diameter of can (m)

D_1 Diameter of rollers in single can simulator (m)

D_T Time to destroy 90% of heat labile factor at a constant temperature T (min)

E_c Emissivity of can (n.d.)

E_f Emissivity of flame (n.d.)

$F_{c\,f}$ View factor from can to flame (n.d.)

$F_{f\,c}$ View factor from flame to can (n.d.)

F_o Sterilisation time equivalent at $T = 121 \cdot 1\,°C$ with $z = 10\,°C$, also, F-value

Gr   Grashof number. $(Gr = \beta g \Delta T D^3 \, \rho g^2 / \mu g^2)$

H Headspace in the can (m)

h_c External heat transfer coefficient (J/s m^2K)

h_i Internal heat transfer coefficient (J/s m^2K)

IS Integrated sterilising value (min)

k_c Thermal conductivity of can (W/m K)

k_g Thermal conductivity of combustion gases (W/m K)

k_1 Thermal conductivity of food (W/m K)

L Height or length of can (m)

M Mass of food in can (kg)

μ_b Fluid viscosity at average bulk temperature (Pa s)

μ_g Viscosity of combustion gases (Pa s)

μ_1 Viscosity of liquid in can (Pa s)

μ_p Viscosity of food (Pa s)

μ_w Fluid viscosity at average temperature of can wall (Pa s)

Nu   External Nusselt No. $(Nu = h_c D / k_g)$

Nu_i Internal Nusselt No. $(Nu_i = h_i D / k_1)$

Pr_i Internal Prandtl No. $(Pr_i = \mu_1 C / k_1)$

Pr_s Crossflow Prandtl No. $(Pr_s = \mu_g C_{pg} / k_g)$

π $3 \cdot 14159$

Q_a Heat accumulation in the can (J/h)

Q_c Convective heat transfer rate to can (J/h)

Q_e Radiative energy emitted by can (J/h)

Q_i Radiative energy incident on can (J/h)

Q_r Radiative energy reflected by can (J/h)

Q_R Net rate of radiative heat transfer to the can (J/h)

Q_T Total heat transfer rate to can (J/h)

Re   Rotating Reynolds No. $(Re = D v_p \rho_g / \mu_g)$

Re_i	Internal Reynolds No. $(Re = D_1(\pi D\omega)\rho_l/\mu_l)$
Re_s	Crossflow Reynolds No. $(Re_s = Dv_s\rho_p/\mu_g)$
Δr_c	Thickness of the wall (m)
ρ_c	Reflectivity of can (n.d.)
ρ_l	Density of liquid in can (kg/m^3)
T	Temperature (K)
T_f	Theoretical flame temperature (K)
T_g	Temperature of combustion gases (K)
T_l	Temperature of liquid in can (K)
T_o	Initial temperature in the can (K)
T_{so}	External surface can temperature (K)
t	Time, units variable
τ_c	Transmissivity of can (n.d.)
σ	Boltzmann constant $(5 \cdot 14 \times 10^{-10} \text{J/s m}^2\text{K}^4)$
U	Overall heat transfer coefficient (J/s m^{2}K)
v_p	Peripheral velocity (m/s)
v_s	Gas crossflow velocity (m/s)
ω	Rotational speed of can (rev/s)
z	Temperature change to cause change in heat resistance by a factor (°C)
β	Coefficient of thermal expansion, $1/T$ for an ideal gas (K^{-1})
ρ_g	Density of combustion gases (kg/m^3)

1. INTRODUCTION

Flame sterilisation of foods is a French invention which was patented in 1957–58 by Cheftel and Beauvais.[1-4] Published results of considerable research on development and application followed from CSIRO in Australia and from the University of California (Davis, California USA). While additional work continued in Europe,[5-8] studies on heat transfer in flame were conducted at various facilities in Australia,[9] Brazil,[10] Canada,[11] Germany[12] and the United States.[13-18]

Flame sterilisation of canned foods is done by direct application of flame at atmospheric pressure, in an open processing vessel. The temperature difference between container and flame is up to 1700°C, as indicated by Casimir et al.[19] This temperature difference makes heating of the containers very rapid. Heat distribution in the containers is enhanced by natural rotation of cans as they travel through the cooker. For larger cans, and viscous foods, natural rotation can be augmented with water cooled, variable-speed belts, or, as in the Australian adaptation,[20,21] agitation is enhanced by periodically reversing direction of can travel using variable-speed shuttle bars. Adequate agitation is critical for effective heat distribution. Therefore, the only foods that are suitable for flame sterilisation are those which can be agitated by the rotation of the cans. Suitable foods are liquids, particulates in liquids and vacuum packed particulates. Foods which cannot be agitated by rotation of cans, such as thick purées, and solid packed meats, are not suitable for flame processing because by their nature, they do not transfer and distribute heat from the heated surface rapidly enough. Thus, overheating, burn-on or buckling (permanent distortion) of the can results.

Flame sterilisation does not damage lithography on the cans, and Noh *et al.*[16,17] indicated that flame pasteurisation of carbonated and other beverages in aluminium cans is also feasible.

There are several reviews and overviews published on flame processing.[22-27] Some make original contributions, whereas others only summarise published information related to the process. After years of work with the Steriflamme ™ concept, the material to follow will emphasise the French flame steriliser. The purpose of the chapter is to familiarise the novice, as well as offer an update on work done in recent years, including both published and unpublished information on machine design, heat transfer, packing styles, processing, monitoring, and economic considerations.

2. DESCRIPTION OF MACHINE DESIGN

2.1. Flame Steriliser
The sealed cans are conveyed to the flame steriliser which accommodates both heating and cooling operations within a single frame, as illustrated in Fig. 1. The inset on the left shows distribution of cans on the infeed separator before entering the preheater. In this example, the infeed separator would run at approximately four times the speed of the conveyor in the flame sterliser. The inset on the right shows the position of the can on the tracks over the flame.

Suitable packs for flame sterilisation are generally the same style used for conventional steam/water processing. Conventional packs can be homogeneous liquid products or particulate foods packed in a fluid medium. These typically consist of approximately two parts particulates and one part covering liquid. Although the packs may vary in specifics of

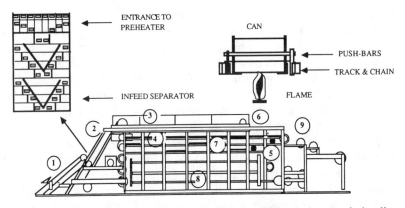

FIG. 1. Schematic drawing of the French flame steriliser. 1, Automatic loading; 2, distribution of cans; 3, preheater with atmospheric steam; 4, riser section to raise can temperature to processing level; 5, automatic temperature recorders/monitors; 6, burners; 7, holding section for maintaining processing temperature; 8, cooling section; 9, can discharge to outlet conveyor. Inset on left shows the can separator. Inset on right is a view of the can directly positioned over the flame.

formulation[28] to make them more suitable for high temperature-short time (HTST) processing in flame, the general pack composition, or fill, is comparable to that used in conventional steam and water processes. The liquid portion of the pack serves as heat transfer medium between the heat source, which is the flame in this application, and the food particulates in the can.

The flame steriliser, with the exception of the preheater, is operated as an open unit. All stages of the process are carried out at atmospheric pressure as described by Thomas.[29] This feature makes the cans in the machine accessible during processing. It is an asset because cans may be removed to verify temperature, and leaky, defective cans are visible and can be eliminated before they cause problems, without interruption of the process. Since the cooker/cooler operates without over-pressure, the cans act as their own pressure vessels during HTST processing of either acid or low acid foods. Proper vacuum, headspace, good closures and proper can-end configuration are critical and generally adequate to maintain container integrity during heating.[30-32]

Machine size and design depend on the range of process times, temperatures and can sizes a processor plans to use. It is common that different foods with different processing requirements are simultaneously flame sterilised in the same cooker/cooler. A single unit will also accommodate several can sizes within the ranges of 114–230 g, or 454–700 g capacities.[7] Separation of products/can sizes is maintained by feeding the cans into and processing them in separate *lanes* of the flame steriliser. Agitation of cans is accomplished by rolling the cans over *tracks*, pushed by bars strategically spaced on conveyor chains. Can rotation is built into the machine, determining, in part, the size of a commercial unit.

To design the machine, maximum can dimension, production rate, minutes of processing per section, and necessary can rotation need to be determined. The parameters are then used as follows: The circumference of the can is the distance travelled in one rotation of the can. Multiplying this distance by the required minimum can *rpm*, will give length of track needed per minute of processing. Then, multiplying distance/minute by the number of minutes needed to achieve processing temperature (time in *riser*), one approximates the maximum length of the machine. *Preheating, holding* and *cooling* sections are built in proportion to the *riser*. If the riser is 2 min long per pass, the other sections will have to be multiples of 2 min (e.g. 2, 4, or 6 min, or one, two, or three passes long).

Compactness of the machine is achieved by stacking. *Preheating* is on top and may have several passes within the steam enclosure. The *riser* is below the preheater, followed by the *holding*, and then the *cooling sections*. Due to limited floor space, the length of machine can be reduced in half, and the sections stacked. Thus by doubling the number of levels in the machine, significant savings in floor space result.

Production speed depends on time/section, diameter of can, and number of *lanes*. In existing commercial machine design, the *push-bars* are spaced (pitched) 1·5 × can diameter apart. Dividing distance of travel/min by the *pitch* will give the maximum number of cans/min in one *lane*, which divided into desired production/min will indicate the number of *lanes* the machine will require for production.

Processing of acid and low acid foods in larger can sizes (> 1000 g) is feasible[31] but may require costly, heavier can structure and expandable can end design for low acid foods, and in general, the augmentation of agitation in critical phases of the process. For large cans, and for viscous foods, the necessary natural can rotation could require a prohibitively large processing unit. When such is the case, water cooled variable-speed belts are used in both the *riser* section and the initial phase of the cooling, to enhance mixing and heat distribution in the canned food. In the Australian flame sterilisers, agitation is similarly enhanced using variable-speed shuttle bars.[20,21] The two sections of forced agitation are critical to achieving rapid heating and cooling, which are necessary for gaining the benefits of HTST attributed to flame sterilisation. However for < 500 g can sizes and most foods, natural can rotation of 30–50 rpm is adequate,[31–33] and machine size remains reasonable without the use of variable-speed belts.

2.1.1. Advantages of Flame Sterilisation

Both processor and consumer related advantages are well described in the literature. For the processor, the openness and accessibility of the flame steriliser unit minimises down-time and loss of production caused by unavoidable jams.[6,8,23] Similar jams in continuous steam processing vessels could cost several days of production and the loss of cans that were in the retort when the process stopped.

The flame sterilisation process is continuous with very high production rates. In several canneries, one flame steriliser had the same out-put as 2–3 rotary pressure cookers processing comparable size cans.[6,34,35]. HTST flame processing also improved retention (higher drained weights) in most products tested.[34–37]

The flame steriliser is also flexible. It can process a range of can sizes and can simultaneously process several different products. This capability exists for hydrostatic retorts, also. However, unlike in a hydrostatic retort each product in the flame steriliser can receive a different heat treatment. While the conveyor speed is set for the longest heating product, flame intensity (temperature) is varied independently in each lane. It would not be unusual to see quality packs of whole peeled tomatoes, peach halves and fruit cocktail flame sterilised in one machine at the same time.

For the consumer, HTST flame sterilised foods retain more fresh-like qualities and nutritive value. Flame processed tomatoes retained more vitamin C than those that were retorted.[34] Better retentions of texture, freshness of flavour and colour have been reported for various fruits and vegetables, also.[34–36,38,39] Even the heat sensitive dairy products were better flame sterilised than retorted, but not as good as when aseptically processed.[40] Enzyme inactivation was often the limiting factor in HTST flame sterilisation of fruits, especially whole, stone-fruits.[39,41] The slower rate of enzyme inactivation and potential regeneration of some enzyme systems in acid foods must not be overlooked.

2.1.2. Disadvantages of Flame Sterilisation

The primary disadvantage of flame processing is its limited ability to process all types of foods and container sizes. As indicated before, very viscous and solid packed foods are not suitable for flame sterilisation. The food must be fluid enough to be agitated by the rotation of the cans rolling over the flame. Although agitation can be enhanced in flame heating, variable-speed belts detract from the simplicity of the machine.

The flame process is best suited to containers less than 1000 g. Larger cans require extra heavy lid construction and special expansion ring design to help them resist the pressures developed during flame heating. Currently available small (< 1000 g capacity) conventionally used containers satisfy requirements of the flame process. However, the conventional supply may soon become inadequate if the trend in reducing plate thickness of cans and ends continues. Unlike in steam retorts, where steam pressure can be used to balance internal can pressures developed during heating,[42] the can is its own pressure vessel in flame sterilisation, and newer, lighter cans may not always have the strength demanded by the process.[7,30] Other trends in new container design do not favour flame processing, either. Flame heating is most effective for metal containers, and it is unlikely that it will be suitable for processing composite or plastic cans.

2.2. High Vacuum Flame Steriliser

The high vacuum flame steriliser (HVFS) unit uses the same flame steriliser described previously. The difference between the processes is the style of pack, vacuum versus brine, and the procedures involved in their preparation.

Vacuum packed foods for HVFS are particulate foods packed under reduced internal can pressure (12–260 mm Hg). The packs consist of 90–100% food and 10–0% liquid added to provide the source of steam necessary for proper processing.[43,44] On heating, the small amount of liquid, either added or released by the food, is turned into steam which is used to remove (flush) non-condensible gases from the packed food and is the heat transfer medium between the food and the heating flame. Since air is an insulator, specified vacuum in the can is essential for delivery of the required process.[45]

Work on the development of HVFS started shortly after the invention and commercial acceptance of the flame steriliser.[46,47] In the experimental design, the primary addition to the flame steriliser was a flame deaeration system. The general concept is illustrated in Fig. 2. Cans filled with food had the lid attached in a loose first operation. The lid was clinched loose enough to allow venting of the can while in the deaerator. In a later modification, only an angular deaerator was used, where the cans with clinched on lids were rotated at an angle while traversing over the hot cone of flames. Boiling of liquid began within 20–40s in hot filled cans of particulates.

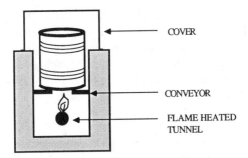

FIG. 2. Schematic view of the vertical flame deaeration unit. The cans are heated while being conveyed without agitation in a flame heated tunnel. Steam generated during heating is used to flush non-condensable gases from the can. Lid on the cans is attached by loose first operation to allow venting. When the venting process is completed, the cans are sealed.

Rotation of the can in the angular deaeration phase permitted use of a hotter flame, and it was assumed that the food would require shorter time to achieve air removal. The deaeration system with both angular and vertical components was found to be the most effective. A photograph of the system is shown by Heil et al.[48] Most of the developmental work and semi-commercial test samples were done on the pictured system.[43,44,49-51]

Leonard et al.[51] found that application of mechanical vacuum instead of flame deaeration produced comparable results in sliced peaches, pears and diced fruit. The concept was further developed and evaluated. It was found that mechanical vacuum deaeration was a good way to produce high vacuum packed foods.[52,53] The procedure required that the foods be hot filled. This was accomplished by blanching, necessary for most vacuum packed foods, to minimise product shrinkage in the can. Product shrinkage occurs in conventional packs, also, but in these, the liquid covering masks the degree of product shrinkage and collapse which results from softening and degassing of the particulates. For mechanical deaeration, good control of fill temperature and of the vacuum level applied to the food is required. Dwell time of 3 s under vacuum was adequate when the vacuum level was high enough to cause flashing of the hot liquids packed with the food.[53]

Parameters of proper mechanical vacuum deaeration can be controlled with ease in commercial application. The proposed commercial unit in which mechanical deaeration can be accomplished is a modified vacuum closing machine. Modification of the vacuum closing machine involves the addition of a vacuumised conveyor tunnel and placement of the transfer valve at the beginning of the tunnel. The size of the tunnel is designed to be proportional to production speed and allow the cans the required minimum 3 s dwell under vacuum before they are sealed in the closing unit. The sealed cans are then discharged either through a second self sealing transfer valve or looped back and exit parallel on the inlet end. The design is fast and adequately compact for commercial use.

2.2.1. Advantages of HVFS

Reduced pressure in the can makes high vacuum packed low acid foods ideal for flame sterilisation, where the cans are heated at atmospheric pressure, rather than under significant over-pressure, as in the rotary steam cookers. To avoid crushing vacuum packed cans in a conventional pressure process, cans must be preheated to near the boiling point of water i.e. atmospheric pressure. This means additional equipment and heat treatment which can detract from quality of heat sensitive foods.

HVFS offers all the advantages of flame sterilisation to the processor. In addition, by omitting the liquid covering which is nearly 30% of the pack, energy needed to heat, handle and transport the conventionally added liquid portion of the pack can be saved.[27,43] Omission of the liquid covering also allows reduction, if not omission, of salt in vegetable and sugar in fruit packs. The processor can choose to either pack more food in the conventional can size, or fit the conventional amount of food into a smaller container; thus, some savings in container cost may also be realised.

Benefits to consumers are numerous.[54] The liquid portion of conventional packs dilutes natural flavour. To enhance flavour, addition of salt and/or sugar is important in conventional packs. In HVFS packs, little or no liquid is added to the natural food whereby its flavour is not diluted, and addition of flavour enhancers is not necessary. Thus, in this age of health and diet consciousness, consumers could finally get foods processed without added salt and/or sugar, without sacrificing natural flavour.[27,43,44,48,51] As a result of HTST processing, retentions of texture,[49,55] colour,[43,44,48-50] flavour[49] and nutritive value[50] of foods are also increased. Work done on nutrient evaluation is limited but conclusive. Seet et al.[50] showed the influence of HTST on retention of heat labile thiamin. Variations in retentions of heat tolerant niacin and riboflavin were primarily due to dilution, or to differences in amounts of water added to the packs. The results are summarised in Table 1. The data are shown as measured and as adjusted for dilution in the packs.

TABLE 1
COMPARISON[a] OF PERCENTAGE RETENTION OF VITAMINS IN CANNED WHITE TUNA

Vitamin	Measured		Adjusted	
	HVFS	Retort	HVFS	Retort
Thiamin	55·27	16·03	70·72	27·11
Riboflavin	85·03	68·85	100·00	100·00
Niacin	86·70	70·40	100·00	100·00

[a] Based on precooked, ready to eat flesh vitamin contents, when processed by HVFS or in a still retort. Measured percentages[27] and values after adjustment for dilution in the packs[50] are given.

2.2.2. Disadvantages of HVFS

High vacuum in the cans requires base-plate weight and temper that will resist panelling and potential damage from stacking. Most current conventionally used consumer-size cans are adequately resistant to internal high vacuum assisted damage. However, the trend of using reduced base-plate weights in can manufacturing is not in favour of either conventional vacuum or HVFS packaging. While mechanical deaeration is applicable to other packaging materials (composite and plastic cans), these may not be suitable for flame sterilisation. Also, several of the foods tested (e.g. mushrooms and olives) did not vacuum pack successfully.[27]

2.3. Fuel

Natural gas, propane and butane are the most commonly used fuels in flame sterilisers. The cans are positioned just above the blue cones of the flames. In this direct flame heating, processing temperatures in cans are achieved in one to three minutes, depending on product, can size, packing style, can rotation and flame intensity. In the original burner design, only the pressure of the combustible gas is regulated. Air is supplied from the surroundings at the burner surface and by aspiration as the jet of gas enters the mixer. Burner capacity will generally depend on pressure and on size of orifice through which the fuel is injected. In subsequent sections, this type of burner will be called a passive burner, since the combustion mixture is not actively controlled.

The flame steriliser uses roughly the same amount of fuel, whether it is operated filled with cans or empty. Therefore, running the machine filled to capacity can significantly reduce unit cost of processing. Energy considerations and equipment modifications to improve efficiency will be discussed in more detail in Section 6.

2.4. Monitoring

Flame sterilisation processes for low acid foods are monitored as required by law.[45] As minimum requirement, cans are removed after the *riser* and *holding sections*, and their temperature measured and recorded manually, at no greater than 15 min intervals. Alternatively, can surface temperatures may be monitored and recorded automatically using suitable temperature measuring devices built into the machine after the *riser* and *holding sections*. Monitoring will be discussed in greater detail in Section 5.

3. HEAT TRANSFER AND DISTRIBUTION

When processing in steam, the can is completely surrounded by a homogeneous heating medium of known thermal properties. With well defined boundaries and parameters, heat transfer has been successfully modelled and heat transfer coefficients for model liquids and food products in axially rotated cans determined. For sucrose solution in cans rotated (80–420 rpm) in atmospheric steam, $\pm 33\%$ correlation for overall heat transfer coefficient was obtained by least square analysis:[56]

$$\mathrm{Nu_i} = h_i \frac{D}{k_1} = 0 \cdot 17 \, \mathrm{Re}^{0 \cdot 52} \mathrm{Pr}_i^{1/3} \left(\frac{L}{H} \right)^{1/3} \tag{1}$$

where $\mathrm{Re} = D_1(\pi D \omega)\rho/\mu$, with values of 600–250 000. For models of water and silicone oils[57] the correlation was expressed as:

$$\mathrm{Nu_i} = U \frac{D}{k_1} = 0 \cdot 434 \, \mathrm{Re}^{0 \cdot 571} \mathrm{Pr}_i^{0 \cdot 278} \left(\frac{L}{D} \right)^{0 \cdot 356} \left(\frac{\mu_b}{\mu_w} \right)^{0 \cdot 154} \tag{2}$$

It covered the following ranges of dimensionless parameters: Re of 12–44 000; Pr_i of 2·2–2300; L/D of 1·11–1·61; and μ_b/μ_w of 1·22–1·79.

Heat transfer in commercial flame heating is rather complicated:

(i) The can is heated in a non-homogeneous environment, where the applied flame is only 0·4–0·95 cm wide, positioned at the centre at right angles to the axis of the rotating can (inset, Fig. 1).

(ii) The method of heat transfer is complex. It involves heating by all three modes: conduction, convection and radiation.

Studies on heat transfer in flame ranged from evaluation of heating curves[12] to the determination of internal, external and overall heat transfer coefficients. For modelling studies, conduction has been assumed to be not significant because direct contact with flame occurs only at a very limited region on the can surface.

When using eqn (1), the calculated internal heat transfer coefficients for a range of model solution viscosities averaged 47% of experimental values. Changing the coefficient from 0·17 to 0·37 was recommended.[14] Later, the correlation equation for internal heat transfer coefficient was further modified:[10]

$$Nu = 0.433 \, Re_i^{0.56} \, Pr_i^{0.60} \, Re^{-0.68} \tag{3}$$

Unless indicated differently, passive burners were used in this and in most other experiments.[10,12,14,18]

The external heat transfer coefficient was indicated to be the limiting factor in heat transfer in flame, regardless of the type of burner used. Reported values ranged from a predicted value[15] of 26·38 J/m²s K to a measured maximum[18] of 57 J/m²s K. Finally, the external heat transfer coefficient (h_c) was actually shown to be the limiting factor, where overall heat transfer coefficients (U) approached the values of h_c as the internal heat transfer coefficient progressively increased ($\gg 2000$ J/s m²K) during flame heating.[17]

In an attempt to model heat transfer in flame, Peralta–Rodriguez and Merson[13,15] used a rectangular burner, the same size as the can. The burner was specifically designed and built for the test by Flynn Burner Corp. (New Rochelle, NY). In the development of the flame heating model, the following assumptions were made:

(i) Can surface temperature, T_{so}, is constant.
(ii) Can contents are perfectly mixed.
(iii) Flow of combustion gases around the can is uniform, and recombination or deposition of combustion products on the can surface is absent.
(iv) Ambient air is not incorporated with combustion gases near the can.
(v) External heat transfer coefficient is constant for the cylindrical part of the can.
(vi) Heat is not transferred through the can ends.
(vii) Flame height remains uniform.
(viii) Radiation from the hot gases leaving the flame may be neglected, assuming radiation only between can, flame and surroundings.

   (ix) Flame temperature at the can surface is the same as calculated from adiabatic combustion, assuming dissociation of combustion components.

Based on the above assumptions, rate of total heat transfer (Q_T) to the can became the sum of both rates of convective (Q_C) and of radiative (Q_R) heat transfer.

Rate of convective heat transfer was described by eqn (4):

$$Q_c = h_c A_c (T_g - T_{so}) \tag{4}$$

The correlation[58] for external heat transfer coefficient (h_c), with and without crossflow, with Reynolds numbers ranging from 2000 to 45 000, and Prandtl numbers from 0·6 to 15, was given by:

$$\mathrm{Nu} = 0\cdot135 \, [(0\cdot5 \, \mathrm{Re}^2 + \mathrm{Re}_s^2 + \mathrm{Gr}) \, \mathrm{Pr}_s]^{1/3} \tag{5}$$

The net radiative heat transfer (Q_R) was the sum of radiant energy (Q_i) reaching the can, less the energies emitted (Q_e) and reflected (Q_r) by the can. They were defined as follows:

$$Q_i = E_f A_f F_{f-c} \sigma T_f^4 \tag{6}$$

$$Q_e = E_c A_c \sigma T_{so}^4 \tag{7}$$

$$Q_r = \rho_c E_f A_f F_{f-c} \sigma T_f^4 \tag{8}$$

Based on reciprocity, $A_f F_{f-c} = A_c F_{c-f}$. As the sum of absorptivity (α_c), reflectivity (ρ_c) and transmissivity (τ_c) is unity, and transmissivity of a metal can is zero, $\rho_c = 1 - \alpha_c$. Since $\alpha_c = E_c$, radiative heat transfer to the can was given:

$$Q_R = \alpha_c A_c \sigma (E_f F_{c-f} T_f^4 - T_{so}^4) \tag{9}$$

The overall heat transfer coefficient was then defined as:

$$U = \cfrac{1}{\left[\dfrac{1}{h_c} + \dfrac{A_c \Delta r_c}{k_c A_{c1m}} + \dfrac{A_c}{h_i A_i} \right]} \tag{10}$$

Equations (4), (9) and (10) were combined, and the equation for rate of heat transfer through the can wall was obtained:

$$Q_T = A_c U (T_g - T_1) + \frac{U \alpha_c A_c \sigma}{h_c} (E_f F_{c-f} T_f^4 - T_{so}^4) \tag{11}$$

The rate of heat accumulation (Q_a) in the can was modelled with the assumption that the contents were perfectly mixed. Since the experiment was designed to eliminate heat loss from the can, $Q_a = Q_T$ was assumed, leading to the following expression:

$$Q_a = MC\frac{dT_1}{dt} = A_cU(T_g - T_1) \tag{12}$$

After integration and definition of initial conditions, where $t = 0$ and $T_1 = T_o$, and assuming that $(E_fF_{c-f}T_f^4 - T_{so}^4)$ is constant, the equation for predicting time/temperature history for liquids in flame heated cans was obtained:

$$\left(\frac{T_1 - T_g}{T_o - T_g}\right) = \exp\left(-\frac{A_cU}{MC}t\right)$$
$$-\frac{\alpha_c\sigma(E_fF_{c-f}T_f^4 - T_{so}^4)}{h_c(T_g - T_o)}\left[1 - \exp\left(-\frac{A_cU}{MC}t\right)\right] \tag{13}$$

The model was subsequently tested and verified.[13] Ratios of predicted to experimental heating rates were near unity for experiments evaluating the effects of combustion gas crossflow velocity (Re > 310) and separation between can and burner (at less than 3·5 cm). Predicted values for variations in can rotation and model fluid viscosities were consistently higher, giving a ratio of 1·25–1·67 over experimental measurements. Overall, the results indicated that approximately 3–10% of heating rate was from radiation. This value is expected to vary with spectral properties of external can enamels.

Gas flow rate and can-burner separation were the most important variables to control. Can rotational speed was not important with liquids, using 1·26 cm headspace. However, the authors[13,15] indicated that their work needs to be expanded to conditions which simulate the commercial burner configuration, can spacing and pack styles (the incorporation of particulates).

Need for additional work is clear. For example, the above results agree with others, that can rotation was not important in model systems; yet, microbiological evaluation of food systems and processing studies showed the need for defining the optimum[31-33] can rotation. Similarly, the role of headspace was not identified to be critical whereas it is in actual processing of foods.

Discrepancies in results on can-to-burner distance also exist. Distances from 0·5 cm to 4·0 cm[16] have been reported as optimum. Definition of optimum distance corresponds to maximum heating efficiency. This efficiency will vary with burner types and capacity. Since burner type and capacity used in the modelling experiments varied, disagreements in results can be expected. However, the value of defining parameters, such as combustion gas flow velocities, etc. which may affect optimum can-to-burner distance, is obvious.

4. PROCESS DESIGN

4.1. Description of Process

In flame heating, unlike heating in steam or retorting, the cans never approach the temperature of the heating medium. Typically, with a given burner setting, can temperature will increase linearly, by the same amount, regardless of the initial can temperature. To avoid process deviations, it is critical that cans have approximately the same initial temperature as they enter the flame heating section.

After the cans are properly distributed on the feed conveyor (Fig. 1), they enter the preheating section. The preheater is an enclosed space filled with steam at atmospheric pressure ($99 \pm 1\,°C$). The preheating period is typically 3–6 min. Its purpose is to heat the incoming cans to uniform initial temperature ($95 \pm 1\,°C$), regardless of the original variations in can-filling temperatures. The targeted $95 \pm 1\,°C$ is rapidly achieved, as shown in Fig. 3. Due to the logarithmic temperature difference relationship between heating

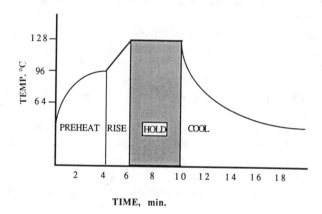

FIG. 3. Typical heating curve in flame processing. Shaded area of 'HOLD' represents a process, calculated in terms of time and temperature required for commercial sterilisation of the canned food.

and heated media, aiming for higher initial temperature in the preheater is not efficient in terms of either time or energy.

After preheating, the cans roll to the riser. In this section, the cans roll over a continuous ribbon of gas flame and are heated to designated sterilisation/processing temperature. According to process design, this internal can temperature may be up to 140°C, depending on the product and packing style. Sterilisation temperatures are typically achieved in 1–3 min.

After the cans are heated to processing temperature, they enter the holding section. Here, the burners are intermittently spaced, and the flames are lowered in intensity. Their purpose is to maintain in-can processing temperature; prevent heat loss, and supply the heat still being absorbed by the particulate food inside the cans.

After holding, the cans roll to the cooling levels where they are cooled under sprays of chlorinated water. The tracks in the cooling section are shallow troughs which collect and retain a 1–2 cm deep layer of water to aid the cooling process. Depth of water varies with can size and is regulated to avoid floating of slack filled or vacuum packed cans. Continuous rotation of cans maximises cooling rate. Air blowers have been used to increase cooling rates by evaporative cooling of cans.

4.2. Processing Variables

4.2.1. Time

As indicated earlier, the flame steriliser is designed for maximum intended process time, temperature, can size, and minimum required can rotation. Time in each section is then regulated by the speed of conveyor-chains which support the push-bars. As in other methods of processing (e.g. rotary pressure cookers and aseptic systems), the faster the cans (or food) are moved through the process, the more agitation they may receive at the expense of residence time in the heating section. Varying time by simple adjustment of conveyor speed is one way of changing process temperatures with or without adjusting flame intensity. When process time is minimised for a given temperature, heating rate (burner output) usually must be maximised. With extreme settings, however, the process becomes less tolerant to deviations in critical parameters such as headspace, fill, consistency, etc.

Since flame sterilisation is a precision process, adherence to scheduled process parameters is more important in flame processing than it is in other methods of in-container processing. For example, increased agitation from faster conveyor speed should not be expected to compensate for shorter time of heating. Similarly, longer time of heating from slower conveyor speed will cause decreased agitation. In either event, one should expect a serious process deviation because the change in one processing parameter will not automatically compensate for the change caused in the other.[45]

4.2.2. Temperature

Flame intensity determines heating rate, and time in the heating section determines heat accumulation (temperature) in the can. Although the various models of heat transfer did not attribute much importance to can rotation, it will affect heat distribution and uptake of heat by the particulates within the can. Conveyor speed, flame intensity and can rotation are critical[59] because changing any one parameter will affect uniformity of heating and final temperature in the food.

4.2.3. Can Rotation

As indicated, can rotation affects distribution of heat within the can. It is less important when processing low viscosity liquids where internal heat transfer is close to ideal convection and becomes increasingly more important when particulates are present in the liquid, or when the liquid is viscous (e.g. fruit nectars or tomato juice). Can rotation helps to reduce resistance to heat transfer at the inner surface of the can wall and on the surface of the particulates.

4.2.4. Critical Factors

If not adequately controlled, critical factors will affect the satisfactory delivery of a scheduled process. Their importance is well documented for all agitating processes.[45] Some of these critical factors are consistency, headspace, particulates-to-liquid fill ratio, particulate size and integrity, and vacuum level. Consistency was one factor already indicated in connection with heat transfer. It should be specified and controlled within tolerance. Headspace must be adequate to allow movement of can contents as the cans roll through the process. Fill ratios are important because the liquid constitutes the heat transfer medium. The tight packing of the particulates will affect their rotational mobility in the can. Control of this factor will have an economical impact, also. Dimensions of the particulates affect the surface area for internal heat transfer from liquid into the food. Since solid foods heat by conduction, their dimensions will determine the length of time needed to attain a designated particle centre temperature.[60] Particulate integrity was also shown to be a factor.[32] Damaged food particles not only carry greater microbial contamination, but do not tumble as freely in the rolling can as do whole particles, affecting both heat transfer and distribution. Minimum designated vacuum is critical to processed product safety in vacuum packed foods. Air insulates, and underprocessing results if minimum vacuum requirements are not met. Vacuum is also important to maintaining container seam integrity. In flame processing there is no over-pressure, and the can must become its own pressure vessel. Internal can pressure developed during flame heating is inversely proportional to vacuum in the can. With higher vacuum, less non-condensible gases remain in the can. Then, internal pressure with heating will be closer to vapour pressure of the food, thus, the pressure the can must withstand will be lower.

4.3. Process Determination and Evaluation

Once all critical factors are defined, determination of the process may begin. Knowing the required F_o-value for the process, one calculates the time and temperature needed for its delivery in the holding section of the flame steriliser. That is, the food is to be heated to the required temperature and held at that temperature until sterilisation of the food is completed, as illustrated in Fig. 3. The principle is the same as the one used in aseptic processing. Lethal treatment accumulated in both the rising (heating) and the cooling phases is ignored in the calculation of the process, for added safety.

The open structure of the flame steriliser allows access in all but the preheating sections. One method of verifying a process is by can centre measurements after each phase of processing and from each lane of the machine.[45] Flame intensity and conveyor speed may be adjusted as necessary[8,59] to achieve process lethality when developing a process. However, only authorised personnel are allowed to make adjustments during production. Leonard et al.[32] showed that isolated cans had significantly lower temperature than cans surrounded by others. Therefore, temperature measurements should be made on cans which will register the least heating, i.e. on isolated cans. In machines equipped with any in-line temperature monitoring capability, internal can temperatures for foods must be correlated to surface temperature readings. Based on this correlation, the process limits of the monitors for a given product can be set.

The process should be also evaluated and confirmed using micro-biological methods.[61-63] One of the best microbiological methods is the count reduction procedure developed by Yawger.[64] The procedure involves inoculation of canned foods, and measuring the reduction in viable colony forming units after heat processing of the cans. The lethal heating value of the process is expressed in terms of Integrated Sterilising (*IS*) values:

$$IS = D_T(\log a - \log b) \tag{14}$$

D_T is time in min needed to destroy 90% of the test organisms in the food at temperature T. Values of a and b are the numbers of test organisms in the food before and after heating, respectively. The term ($\log a - \log b$) represents the count or decimal reduction value which can be used comparatively, without determination of D_T, if the suspension or source of microorganism and formulation of food are identical. The count reduction procedure is a microbiological measure of process adequacy. It can be used for transferring process lethality from one proven process (from retorting) to a new process (flame steriliser),[61] by matching the destruction of test organisms. When statistically analysed for dispersion, the count reduction data can also be used to optimise headspace, viscosity, fill and can rotation via changing the critical conditions to obtain minimum standard deviation from the mean[61,64] *IS* or count reduction values.

Methods of inoculation include direct mixing of test organisms into the liquid, inserting sealed capillary tubes with calibrated inoculum into the food, injecting particulates directly with inoculum, and/or preparing inoculated synthetic particulates.

Calibrated inoculum in capillaries is a suspension of spores of known concentration and specified heat resistance. The advantage of this method is that placement of inoculum is well defined and potential of contamination during handling is minimised. The disadvantage is that the spores are heated without being exposed to the food environment.

Preparation of the synthetic food particles is described by NCA.[65] A known number of spores is mixed into the puréed particulates. Egg whites are added to the mixture, which is then spread into a layer of desired thickness. It is heated at sub-lethal temperature, only to set (coagulate) the egg whites. The semi-solid preparation is then cut into chunks of desired dimensions and placed with natural particulates into the test cans. The advantage of the method is that the particulates contain a known amount of inoculum, as compared to direct injection where the inoculum tends to leak out of the particulates. The disadvantage of the procedure may be that heat transfer characteristics of the food could be altered, and the inoculum is in a modified food environment.

When the process is designed using only temperature measurements, it should be confirmed using inoculated packs. Procedures are described in adequate detail by NCA.[65]

5. MONITORING

In steam retort processes, monitoring and recording steam temperature is an adequate procedure for documenting that the process was delivered for the scheduled time and temperature.[45] Monitoring the temperature of the flame in flame sterilisers has been done,[21,59] but it is complicated. The environment is highly oxidative for thermocouples and, as shown earlier, a number of factors enter into actual delivery of a process. As a result, rather than monitoring flame temperature, it is preferred that accumulation of heat in cans be measured. Although continuous monitoring is preferred, the minimum requirement is that can centre temperatures be measured at no more than 15 min intervals after rising and holding, from each lane of the flame steriliser.[45] Experience with the dependability of the flame sterilisation concept justifies this relaxed monitoring schedule. However, the potential of fuel quality fluctuation and supporting equipment malfunctioning is always present, and the packing parameters, however slightly, vary constantly. Finding a dependable method for in-line monitoring of can temperatures has been an important and long quest.

Thermocouples, thermistors, temperature sensitive paints, photo-cells and infrared sensors have all been tested for temperature monitoring purposes. Cost, sensitivity, response time and reliability were major factors in selection of a suitable system. While industry used only plain tin containers non-invasive thermocouple systems were developed and optionally used for automatic monitoring in commercial flame sterilisers.[6,66] Some of the thermocouples were constructed from copper–constantan foils, giving a relatively large surface to mass ratio, whereby on contact with the hot can, the thermocouple reached can surface temperature within the time limitations set by production speed. Although the system was automatic, only a sample of the cans processed was actually monitored. The monitoring procedure was non-destructive, and by the number of cans sampled, it was superior to the manual method. The automatic monitoring procedure assumed that contents of cans were perfectly mixed, and that the can end surface temperature was representative of the temperature of the food. Merson et al.[14] showed that heat distribution in flame heated cans is, indeed, nearly uniform, and that the 'cold' region was located at the ends.

Although the above monitoring systems developed for measuring can end temperatures were appropriate, to improve reliability, an invasive method of measuring internal temperature was also designed.[67]

5.1. Copper–Constantan System

The first continuous monitoring device for plain tin cans was developed by Beauvais,[68] and results of its evaluation reported by Leonard *et al.*[69] The device is schematically illustrated in Fig. 4. Three features set the device apart from its predecessors. First, it could be built into the machine as part of the track after rising and holding, where every can rolling through the processing unit is monitored. Second, the steam in the copper tubing is controlled at near processing temperature, thus the speed of monitor response is increased and baseline drifting caused by changing monitor temperature is eliminated. Third, the thermocouple circuit is completed by the plain tin (Sn) surface of the can. The thermocouple-wire-to-can-surface generated signal is cancelled as shown below, and the copper–constantan thermocouple millivolts (mV) are recorded:

$$mV_{Cu-Sn} - mV_{Const.-Sn} = mV_{Cu-Const.} \qquad (15)$$

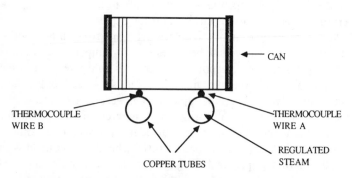

FIG. 4. Schematic presentation of a copper–constantan thermocouple monitoring device. The plain metal body of the can completes the thermocouple circuit. Regulated steam is at near processing temperature for the cans.

Surface temperatures measured on this monitoring device had very good correlation to internal can temperatures ($R^2 = 0.99$). The device is inexpensive and reliable, provided it is properly maintained by regular cleaning of the wire contact surfaces. However, this monitor requires that the plain metal surface of cans complete the thermocouple circuit, and industry since has generally changed to use of tin free steel (TFS), outside enamelled containers. TFS can surface is not electrically conductive because the metal surface is covered with an insulating epoxy/enamel to protect the can from corrosion, thus the monitoring device cannot work with TFS cans.

5.2. Infrared System

The energy measurable in the infrared region of the electromagnetic spectrum is proportional to the emissivity and temperature of the emitting body. Reliability of converting the radiated energy to temperature measurements is very dependent on emissivity and on working temperature range. Plain tin can surfaces have less than 12% emissivity. Also, when compared to successful applications with glowing hot and molten metals, the temperature range of interest in thermal processing is relatively low. Thus, infrared thermometry was unsuitable when monitoring processes using plain tin containers were first considered. While lithographed and enamel coated TFS cans are not electrically conductive, the emissivity of the epoxy/enamel coating is adequately high (> 50%) to make infrared monitoring applicable to these containers.[69]

Physical evaluation of infrared thermometry for monitoring showed good correlation ($R^2 \geq 99\%$) between can surface and centre temperatures. The results were most consistent when can end, rather than body surface temperatures were observed. There are several advantages to monitoring can end surface temperatures:

(i) It is the slowest heating region in flame heating.[14]
(ii) Use of TFS ends can be universal. TFS ends are more economical, and may be used with plain tin can bodies.
(iii) Monitored surface on the end (Fig. 5) can be covered, whole or in part, with nearly 100% emissivity coating which would improve reliability of the monitoring system. It could be used, also, to identify flame processed products. While maximum emissivity has been classically assigned to flat-black materials, infrared thermometry is essentially colour-blind. That is, the monitored region can be colourfully embellished with high emissivity materials, maximising both emissivity and aesthetic appeal.

5.2.1. Microbiological Evaluation of the System

The infrared monitoring system was further evaluated using microbiological procedures. The material to follow is an original contribution, therefore it is sufficiently detailed to provide supporting information.

FIG. 5. Schematic presentation of the infrared monitoring concept. Approximate surface viewed by the monitor is shaded and may represent a design specific for flame sterilised products. It can be painted in colour with high emissivity epoxy/enamels.

The work involved extensive inoculated vacuum packs of cut green beans. The beans were flame sterilised and the processes were monitored using the infrared monitors at the end of rising and holding. The results were analysed for process deviations, and used to set and verify the temperature needed to produce a commercially safe high vacuum flame sterilised (HVFS) product. Compared to the conventional, HVFS packs had vacuum in the can as an additional critical factor, making the vacuum packed food the ultimate challenge for the study.

Can end surface temperatures after the riser and the corresponding spoilage of cans are shown in Fig. 6. The beans were blanched 5 min at

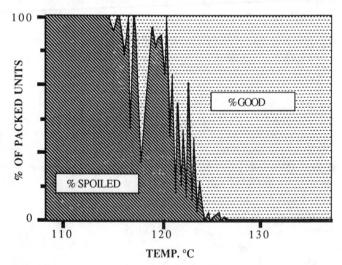

FIG. 6. Results of inoculated pack study with cut green beans.

79·4°C, deaerated, flame heated from 1·2–2·5 min on the riser, and held equal lengths of time in the holding section. Additional holding without flame was also used to complete heat penetration into the food particles. Heat accumulation, total heating time and holding without flame are given in Table 2. Each can was inoculated into the beans with $9·95 \times 10^4$ spores of PA 3679 (Putrefactive Anaerobe 3679), grown according to the procedure of Goldoni et al.[70] Using PA 3679 in place of the spores of the toxin producing *Clostridium botulinum* is a classical procedure. Although PA 3679 spores respond like spores of *C. botulinum*, the former are more heat resistant and do not produce toxin.

TABLE 2
FLAME PROCESS FOR INOCULATED HVFS GREEN BEANS

Process level	Total time flame section (min:s)	Hold without flame (min:s)	Heat accumulation in can[a] (kcal)
1	3:42	1:14	18·27
2	3:00	1:00	16·68
3	2:54	0:58	14·52
4	2:36	0:52	13·00
5	2:12	0:44	11·01
6	1:51	0:37	9·25

[a]Heat accumulation was measured using the procedure described by Leonard et al.[69]

The green beans had pH = 6·5. The 90% reduction time at 115·5°C, i.e. the $D_{115·5°C}$ value of the spores was 2·35 min with $z = 10°C$, meeting the heat resistance specifications set by NCA.[65] The results indicated that cans with 125°C or higher surface temperature with minimum 2 min of heating showed no evidence of spoilage. That is, the process requirements were satisfied by process levels 1, 2, and 3, where no spoilage was experienced.

Subsequently, vacuum packed small potatoes were also used for evaluating the monitoring system. Potatoes were lye peeled and held in 0·1% citric acid solution until canned. Chunks weighing 10–30 g were inoculated with 10^2 spores of PA 3679. An additional surface inoculum of 10^4 was added to each can. The packs were processed but no spoilage was observed above 115·5°C of measured can surface temperatures. The results showed that even with marginal heat treatment, the residual citric acid on the potatoes (pH $\geq$ 5·6) was sufficient to inhibit growth of the surviving test spores. Therefore, it was important that the test be repeated without using citric acid in any phase of potato preparation for canning.

In the second test, a stepped-up inoculation procedure was used. The procedure had the advantages of speed and the smaller number of cans it required. Heat accumulation and flame heating time and holding without flame are given in Table 3. With larger particulates, longer time was needed for the heat to penetrate to the centre of the food.

TABLE 3

FLAME PROCESS FOR INOCULATED HVFS WHOLE POTATOES

Process level	Total time flame section (min:s)	Hold without flame (min:s)	Heat accumulation in can[a] (kcal)
1	6:32	6:32	24·29
2	6:00	6:00	22·30
3	5:00	5:00	18·60
4	4:00	4:00	15·30
5	3:40	3:40	14·87

[a]Heat accumulation was measured using the procedure described by Leonard et al.[69]

The potatoes were injected with 10^2, 10^3 or 10^4 spores per can, with surface inoculation of 10^4 constant for all. The cans were processed, and the results are shown in Fig. 7. Potatoes heat differently from green beans and F-values calculated from monitor readings, instead of temperaturés measured after heating, are given. With standard 10^2 inoculation level in the potatoes, an F_o-value of 10 min was adequate to destroy the test organisms. This F_o-value corresponds to a minimum of 127°C can surface temperature at both stages of monitoring. With inoculum levels of 10^3 and 10^4 spores in the potatoes, the required F_o-values increased to 25 and 30 min, respectively. Process levels 1, 2, and 3 had adequately high F_o-values, since no spoilage from these levels was observed.

The results of a stepped up inoculation procedure can also assist in deciding on margins of safety one should add to a given process. This must be considered, if strict control of critical factors is not feasible. An example is shown in Table 4. Potato size significantly affected the sensed can surface temperatures. With smaller potatoes, the conduction heating surface was

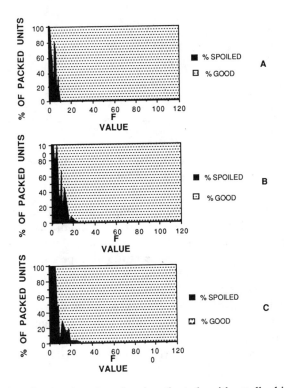

FIG. 7. Results of stepped-up inoculated pack study with small white potatoes. Plot A shows spoilage with 10^2 spores injected into the potatoes. Plots B and C show results for 10^3 and 10^4 inoculation levels, respectively.

significantly greater than with larger potatoes.[60] Therefore, the internal heating medium temperature, as monitored in the machine, showed apparently lesser accumulation of heat with small particulates than with the large. As a result, in setting the minimum monitor alarms, one should test with the smallest particles when processing foods in which uniform particle sizes cannot be assured.

5.2.2. Indications of Process Deviation

First, it was shown that infrared monitoring correlated well to internal can temperatures. This was microbiologically evaluated and verified. The next step was to show whether the monitoring system was adequately sensitive to signal deviations due to malfunctions during the flame process.

TABLE 4

INFLUENCE OF PARTICLE SIZE IN VACUUM PACKED POTATOES ON SURFACE
TEMPERATURES OF THE CANS

Weight range (g)	Temperature °C end of hold[a]	Calculated F_o can end surface
≤ 10	127·4 ± 0·6	19·7 ± 4·1
11–20	129·0 ± 1·0	23·2 ± 5·3
21–30	130·6 ± 1·0	27·5 ± 6·2

[a] Measured with infrared temperature monitor. Particle size is specified by weight.

There were two cases of equipment malfunctioning during the work with the test machine. The first case involved power failure after which the air compressor failed to restart. Loss of air pressure diminished the intensity of the riser flame, and surface temperatures dropped drastically. Figure 8 shows the normal monitor temperature levels for green beans over those recorded while experiencing the process deviation.

The second case involved whole peeled potatoes. The data are shown in Fig. 9. The process deviation was caused by an accidental shut-down of a holding burner. While the burner was out, deviation of surface temperature from the normal run is clear. The data also show that

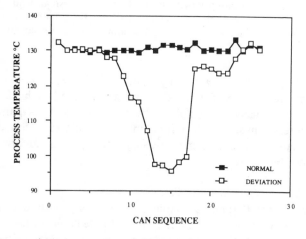

FIG. 8. Process deviation monitored during flame processing of vacuum packed cut green beans. Deviation was caused by loss of air pressure. Temperature history of normal processing condition is superimposed for comparison.

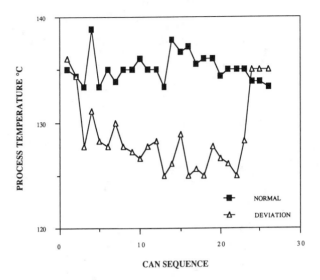

FIG. 9. Process deviation monitored during flame processing of vacuum packed small potatoes. Deviation was caused by shut-down of a holding burner.

processing temperatures for potatoes were significantly higher and more variable than for green beans. Close control of fill and size of particulates in potatoes was not practical.

Infrared thermometry is capable of monitoring the thermal process delivered by the flame steriliser to every can processed. It is a non-contacting method for documenting the process given to each can. With proper computer assistance, it could also support the control of functions responsible for safe flame sterilisation of foods.[71] Compared to other systems of monitoring, reliability and process control potentials of the infrared monitoring system may offset its higher price.

6. ECONOMIC CONSIDERATIONS

Costs of equipment and its operation are important factors in enhancing commercial interest and acceptance of innovative processes.

6.1. Equipment
As shown in Fig. 1, the flame steriliser is a simple machine. Cost estimates[27] for a 200 can/min unit were very competitive with estimates given for

TABLE 5
COMPARISON OF ENERGY EFFICIENCY OF VARIOUS PROCESSING EQUIPMENT

Equipment	Source				
	Casimir[72]	Ferrua and Col[73]	Carroad[43]	Singh[74]	Paulson et al.[11]
Still retort	32–45	—	—	—	40·6
Rotary					
Pressure retort	—	47·7	—	30–40	—
Rotary					
Atm. cooker	—	31·2	29·0	—	—
Flame (French)	27	27·5	—	29–46	26·6–27·6
Flame (CSIRO)	41	—	—	—	—
Pilot HVFS	—	—	41·4	—	—

aseptic, rotary pressure and pouch processing systems. However, what those estimates did not show is that the cost of a 400 or 600 can/min flame steriliser would increase to a smaller extent than the costs of other systems comparably upgraded.

To date, all commercially used flame sterilisers in Europe and America had been built in France. Currency exchange and rates of financing often favoured the cost of flame sterilisers over cost of other processing units.

6.2. Energy Usage

Data on energy usage, in terms of efficiency of flame steriliser and several other processing methods, are compiled in Table 5. Variability of results is caused by differences in operation and modifications of the equipment. In comparing pressure cooker efficiencies, boiler efficiencies used in the calculations varied. Also, evaluation of the flame steriliser often included break periods when no cans were produced. In some cases, energy saving measures were taken with operation of the retorts.

Energy used in the preheater was estimated at 38% of the total energy used in the flame steriliser.[11] In pilot flame sterilisers, significantly less steam was needed when the preheating section contained a layer of condensate in the bottom of the enclosure. Presence of condensate ensured that cans were heated in an atmosphere of saturated steam rather than in dry steam that was produced with the aid of heat from the riser section positioned beneath the preheater. High values shown for the flame steriliser by Singh[74] may be due, in part, to the incidental use of saturated steam during testing.

Casimir[72] used an air blower with the Venturi mixer on the CSIRO flame steriliser, whereas the French flame steriliser used only the passive mixer. Also, energy recovering measures were taken, capturing and utilising the heat escaping through the stacks of the CSIRO unit.[21] This may have significantly increased the efficiency of that flame processing operation.

Considering all the variables involved in evaluating the efficiencies of the various processing units, the results are not totally in disagreement. All of the investigators indicated that much could be done to improve the efficiency of flame processing.

6.3. Potential for Improvements

Paulson et al.[11] recognised the high energy demand of the preheater. They suggested the use of hot water, partially heated with energy recovered from flame heating, for preheating of cans. Although heat transfer rate in water is not comparable to heat transfer in steam, the idea deserves further study.

Similarly, measuring energy demand in the preheater, when properly operated with saturated steam, could use another look.

Recycling of cooling water was indicated as a potential source for energy recovery. It was estimated that 30 litres of water could be heated from 10°C to 90°C in the first minute of cooling of flame sterilised cans, at 192 cans/min production rate. This hot water could be utilised for cleaning or as boiler feedwater. Other energy recovery measures would involve design and installation of energy traps.

6.3.1. Energy Traps

In the Australian machine design, flame heating in the riser was used for producing steam for preheating of cans.[20] Obviously, this feature enhanced the efficiency of the CSIRO flame steriliser (Table 5). Additional recovery of waste heat from the flue[21] was utilised for warming water for washing and rinsing cans and for space heating during the cold season. Potential savings from deliberate combustion-energy recovery measures were estimated at 19·9% of the total energy used for flame sterilisation.[11]

6.3.2. Burner Design

The Australian burner design used a Venturi mixer with a blower to mix air and combustible gases before the mixture reached the burner. This design was superior to the passive system used in the French machines. A small, 15 cm version of the passive mixer is shown in Fig. 10. Burner capacity was regulated by size of orifice (A), and fuel-to-air ratio by the distance (B) between the orifice and mixer.

An ideal burner system was provided by the Flynn Burner Corp. (New Rochelle, NY). The system has replaced the gas jets with burner ribbons. Depth, contour and the needed number of ribbons are calculated for precise control of flame size and shape for any application. Flynn systems are also available for burning liquid fuels (diesel, kerosene, etc.), but these have not been tested yet for application in flame sterilisers. In addition to the unique burner design, the system regulates compressed air, and when adjusted properly, it will mix with a stoichiometric ratio of combustible gas. As a result, fuel is not wasted, but burned completely. The Flynn mixer system used on the University of California pilot flame steriliser is shown in Fig. 11. To reduce demand on compressed air, ambient air was used to augment the air requirement. A comparison of burner efficiencies is given in Table 6. Operating capacity for both burners was near the gas flow rate of 1·59 litres/min. The projection of heat from the burner, i.e. the velocity of hot gases from the Flynn system, was proportional to the pressure of air

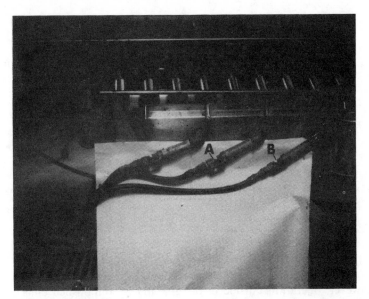

FIG. 10. Picture of a passive burner system. The orifice size (A) determines capacity. The distance from the mixer (B) regulates fuel-to-air ratio.

FIG. 11. Picture of the mixer and controls that were used on the test machine with the Flynn burners. Air pressure (A) is regulated and fuel (B) is aspirated into flow. Air pressure gauges (D) are standard, but the flow tube (C) for measuring fuel consumption rate was an added option.

TABLE 6

COMPARISON[a] OF THE EFFECTIVENESS OF COMPARABLE CAPACITY PASSIVE BURNER
DESIGN WITH A PASSIVE MIXER SYSTEM WITH THE FLYNN DESIGN USING A SYSTEM
OF POSITIVE AIR/FUEL CONTROL

Gas flow rate	Passive burner	Flyn burner
0·44	397	409
1·59	1052	1251
2·22	1532	1567

[a] Based on heating rates (J/s) in the cans at indicated gas flow rates (litres/min), at
0·5 cm can to burner distance.

employed. Operating at designated flow rates for the specific burners, the
Flynn burner was significantly more efficient than the passive burner.
However, operation outside of design capacity diminished efficiency of
both. The burner system used by CSIRO was not available for testing.

6.3.3. Conveyor Design
Energy waste in the preheater has been indicated.[11] The other significant
energy wasting feature in existing flame sterilisers is the conveyor design.
More specifically, the spacing of push-bars.

In the original flame steriliser design, transfer-bars for both pushing and
retaining of cans while being transferred between levels of the flame
steriliser were employed (Fig. 12(A)). The system was probably quite energy
efficient, but it completely denied access to the cans, thus it had to be
modified. Subsequent commercial machine design had pairs of round bars
(Fig. 12(B)) spaced (pitched) 1·5 × can diameter apart. The spacing was
necessary to minimise the influence of adjacent cans (or lack of them) on
rate of heating in the riser section. Even with the above pitch, cans
surrounded by other cans were found to be significantly hotter than those
processed alone.[32]

Recently, a flat push-bar system was laboratory tested (Fig. 12(C)). The
bars were 0·5–0·7 cm thick and approximately as wide as $\frac{1}{2}$–$\frac{2}{3}$ the diameter of
the cans processed. The bars were positioned with lower edges just above,
away from the hot cones of the flames. They were installed on the conveyor
chain, as were the round bars, for easy removal. However, the pitch was
only about one centimetre greater than the diameter of the cans, so that
cans were still accessible for removal and manual monitoring. Cans
processed in a group had the same average temperature as cans that were

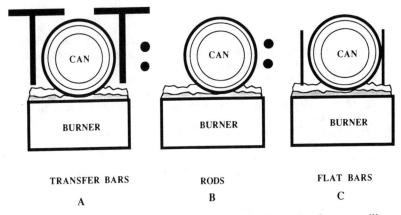

FIG. 12. Schematic presentation of push-bars used in the flame sterilisers. A, originally used transfer bars; B, rod type push-bars used on existing commercial machines; C, laboratory tested flat bars with energy saving potentials.

heated alone. By this new design, future reduction of the pitch is expected to amount to significant savings in fuel efficiency. The savings should be in proportion to the ratio of unused flame (dead-space) eliminated or recovered/utilised in the rising and holding sections.

DEDICATION

This chapter is dedicated to the memory of the late Sherman J. Leonard who devoted the last decade of his life to the advancement of knowledge and understanding of flame processing.

REFERENCES

1. CHEFTEL, H. and BEAUVAIS, M. French Pat., 1957, Oct. 28, 1,540,099.
2. CHEFTEL, H. and BEAUVAIS, M. French Pat., 1958, May 12, 1,164,343.
3. CHEFTEL, H. and BEAUVAIS, M. French Pat., 1958, July 29, 1,180,283.
4. CHEFTEL, H. and BEAUVAIS, M. French Pat., 1958, Dec. 27, 1,189,333.
5. BEAUVAIS, M., THOMAS, G. and CHEFTEL, H. A new method for heat-processing canned foods. *Fd Technol.*, 1961, **13**, 5–8.
6. THOMAS, G. The French Steriflamme and its applications, 1961, private communication.

7. THOMAS, G. Behavior of cans processed in Steriflamme, 1961, private communication.
8. THOMAS, G. Influence of stopping the 'Steriflamme' on the lethal efficiency of the heat treatment effectively applied to the cans being processed, Project Report, Aug. 1971.
9. WU, B.-K. Aspects of heat transfer in flame sterilization of canned milk, M.Sc. thesis, 1971, University of South Wales, Kensington, Australia.
10. TEIXEIRA NETO, R. O. Heat transfer to liquid foods during flame-sterilization. *J. Fd Sci.*, 1982, **47**, 476–81.
11. PAULSON, A. T., VICTOR LO, K. and TUNG, M. A. Conventional steam retort versus direct flame sterilizer in the thermal processing of canned low-acid foods. *Report. Eng. Stat. Res. Inst.*, Ottawa, Ont. K1A 0C6, 1984.
12. PAULUS, K. and OJO, A. Heat transfer during flame-sterilization. *Proc. IV Int. Cong. Fd Sci. and Technol.*, 1974, **IV**, pp. 443–8.
13. PERALTA-RODRIGUEZ, R. D. and MERSON, R. L. Experimental verification of heat transfer model for simulated liquid foods undergoing flame sterilization. *J. Fd Sci.*, 1983, **48**, 726–33.
14. MERSON, R. L., LEONARD, S. J., MEJIA, E. and HEIL, J. R. Temperature distribution and liquid–side heat transfer coefficients in model liquid foods in cans undergoing flame sterilization heating. *J. Fd Proc. Engrg*, 1981, **4**, 85–98.
15. PERALTA-RODRIGUEZ, R. D. and MERSON, R. L. Heat transfer and chemical kinetics during flame sterilization. *AIChE Symp. Series*, 1982, **78**, 58–67.
16. NOH, B. S., HEIL, J. R. and PATINO, H. Effect of processing variables on flame pasteurization of liquids in aluminium cans. *J. Fd Sci.*, 1985, **50**, 1448–52.
17. NOH, B. S., HEIL, J. R. and PATINO, H. Heat transfer study on flame pasteurization of liquids in aluminium cans. *J. Fd Sci.*, 1986, **51**, 715–19.
18. FUJIWARA, M. External heat transfer coefficients during canning by flame sterilisation, M.Sc. thesis, 1975, University of California, Davis.
19. CASIMIR, D. J., HUNTINGTON, J. N. and RUTLEDGE, P. J. Control of the thermal process in flame sterilization. *CSIRO Fd Res. Q.*, 1975, **35**, 63–7.
20. HUNTINGTON, J. N. and CASIMIR, D. J. Design, construction, and operation of reversing-roll pilot-scale flame sterilizer. In *Flame Sterilization. Specialist Courses for the Food Industry*, AIFST–CSIRO, North Ryde, NSW, Australia, No. 2, 1972, pp. 27–32.
21. LEWIS, P. S. and LOHNING, F. M. A commercial scale reversing-roll flame sterilizer. In *Flame Sterilization. Specialist Courses for the Food Industry*, AIFST–CSIRO, North Ryde, NSW, Australia, No. 2, 1972, pp. 42–6.
22. CASIMIR, D. J. Flame sterilization. *CSIRO Fd Res. Q.*, 1975, **35**, 34–9.
23. CASIMIR, D. J. New equipment for the thermal processing of canned foods. In *Flame Sterilization. Specialist Courses for the Food Industry*, AIFST–CSIRO, North Ryde, NSW, Australia, No. 2, 1972, pp. 1–9.
24. LEONARD, S., MERSON, R. L., MARSH, G. L., YORK, G. K., HEIL, J. R. and WOLCOTT, T. K. Flame sterilization of canned foods: An overview. *J. Fd Sci.*, 1975, **40**, 246–9.
25. LEONARD, S. J. and LUH, B. S. Steriflamme heat processing of canned fruits and vegetables. In *Recent Advances in Food Science and Technology* (eds S. M.

Chang *et al.*, Vol. 1, Hua Shiang Yuan Publ. Co., Taipei, Taiwan, 1981, pp. 260–70.

26. RICHARDSON, P. S. Review: Flame sterilization. *Int. J. Fd Sci. and Technol.*, 1987, **22**, 3–14.

27. HEIL, J. R. High vacuum flame sterilization of particulate foods: Improved quality and economics. In *Proc. First Intrntl Conf. on New Innovations in Packaging Technologies and Markets*, 1983, pp. 28–30.

28. CASIMIR, D. J. and LEWIS, P. S. Product formulation specifically for flame sterilization. In *Flame Sterilization. Specialist Courses for the Food Industry*, AIFST–CSIRO, North Ryde, NSW, Australia, No. 2, 1972, pp. 59–61.

29. THOMAS, G. The French Steriflamme and its applications. In *Flame Sterilization. Specialist Courses for the Food Industry*, AIFST–CSIRO, North Ryde, NSW, Australia, No. 2, 1972, pp. 49–53.

30. CASIMIR, D. J. Container requirements for flame sterilization. In *Flame Sterilization. Specialist Courses for the Food Industry*, AIFST–CSIRO, North Ryde, NSW, Australia, No. 2, 1972, pp. 33–5.

31. LEONARD, S., MARSH, G. L., YORK, G. K., MERSON, R. L., HEIL, J. R., WOLCOTT, T. K. and ANSAR, A. Flame sterilization of some tomato products and fruits in 603 × 700 cans. *J. Fd Sci.*, 1976, **41**, 828–32.

32. LEONARD, S., MARSH, G. L., YORK, G. K., HEIL, J. R. and WOLCOTT, T. K. Evaluation of tomato canning practices using flame sterilization. *J. Fd Proc. and Pres.*, 1977, **1**, 313–23.

33. RIEGEL, L. A., JOSEPH, R. L. and GOULD, W. D. Flame sterilization of canned tomatoes. In *1974 Research Progress Reports*, Horticulture series No. 403, Feb. 1974, Department of Horticulture, OSU, Columbus, Ohio.

34. LEONARD, S., MARSH, G. L., MERSON, R. L., YORK, G. K., BUHLERT, J. E., HEIL, J. R. and WOLCOTT, T. K. Chemical, physical and biological aspects of canned whole peeled tomatoes thermally processed by Steriflamme. *J. Fd Sci.*, 1975, **40**, 254–6.

35. LEONARD, S. J., MARSH, G. L., MERSON, R. L., YORK, G. K., BUHLERT, J. E., HEIL, J. R. and WOLCOTT, T. K. Quality evaluation of canned fruit cocktail experimentally processed by Steriflamme. *J. Fd Sci.*, 1975, **40**, 257–8.

36. KLEPETKO, V. G. and LONGWORTH, I. N. Flame sterilization of mushrooms. In *Flame Sterilization. Specialist Courses for the Food Industry*, AIFST–CSIRO, North Ryde, NSW, Australia, No. 2, 1972, pp. 62–4.

37. GELBER, P. Continuous flame sterilization eliminates 5% of product shrinkage. *Fd Proc.*, July 12, 1968.

38. JOSEPH, R. L., RIEGEL, L. A. and GOULD, W. A. Flame sterilization process for canned foods. In *1974 Research progress reports*, Horticulture series No. 403, Feb. 1974, Dept of Hortic., OSU, Columbus, Ohio.

39. LEONARD, S. J., MARSH, G. L., YORK, G. K., HEIL, J. R., WOLCOTT, T. K. and COGGINS, S. Determination of flame sterilization process for apricot halves in 303 × 406 cans. *J. Fd Sci.*, 1976, **41**, 1222–4.

40. KIESEKER, F. G. Methods for sterilization of dairy products. In *Flame Sterilization. Specialist Courses for the Food Industry*, AIFST–CSIRO, North Ryde, NSW, Australia, No. 2, 1972, pp. 54–8.

41. GILLESPY, T. G. and THORPE, R. H. Spin cooking and cooling of canned fruit,

III Flame processing. In *The Fruit and Vegetable Canning and Quick Freezing Research Association, Tech. Mem. No. 60*, 1963.

42. ADAMS, H. W. and OWEN, W. Internal pressure of cans agitated during heating and cooling. *Fd Technol.*, 1972, **24**(7), 28–30.

43. CARROAD, P. A., LEONARD, S. J., HEIL, J. R., WOLCOTT, T. K. and MERSON, R. L. High vacuum flame sterilization: Process concept and energy use analysis. *J. Fd Sci.*, 1980, **45**, 696–9.

44. LEONARD, S. J., HEIL, J. R., CARROAD, P. A., MERSON, R. L. and WOLCOTT, T. K. High vacuum flame sterilized fruits: Storage study on sliced clingstone peaches, sliced Bartlett pears, and diced fruit. *J. Fd Sci.*, 1983, **48**, 1484–91.

45. FPI. *Canned Foods. Principles of Thermal Process Control, Acidification and Container Closure Evaluation*, 4th edn, The Food Processors Institute, Washington, DC, 1982.

46. Anon. Sterilizes vegetables in 1 to 5 min. *Fd Engrg*, July 1973, 69.

47. CASIMIR, D. J., RUTLEDGE, P. J. and HUNTINGTON, J. M. Flame sterilisation of vacuum-packed products. *CSIRO Fd Res. Q.*, 1976, **36**, 25–8.

48. HEIL, J. R., CARROAD, P. A., MERSON, R. L. and LEONARD, S. Development of high vacuum flame processes for sliced peaches and pears. *J. Fd Sci.*, 1983, **48**, 1106–12, 1123.

49. O'MAHONY, M., BUTEAU, L., KLAPMAN-BAKER, K., STAVROS, I., ALFORD, J., LEONARD, S. J., HEIL, J. R. and WOLCOTT, T. K. Sensory evaluation of high vacuum flame sterilized clingstone peaches, using ranking and signal detection measures with minimal cross-sensory interference. *J. Fd Sci.*, 1983, **48**, 1626–31.

50. SEET, S. T., HEIL, J. R., LEONARD, S. J. and BROWN, W. D. High vacuum flame sterilization of canned diced tuna: Preliminary process development and quality evaluation. *J. Fd Sci.*, 1983, **48**, 364–374.

51. LEONARD, S. J., HEIL, J. R., CARROAD, P. A., MERSON, R. L. and WOLCOTT, T. K. High vacuum flame sterilized fruits: Influence of can type on storage stability of vacuum packed peach and pear slices. *J. Fd Sci.*, 1984, **49**, 263–6.

52. HAMBLIN, C. L., HEIL, J. R., BERNHARD, R. A., MERSON, R. L. and PATINO, H. Comparing flame and mechanical deaeration of high vacuum canned green beans and apple slices. *J. Fd Sci.*, 1987, **52**, 425–8.

53. HEIL, J. R., HAMBLIN, C. L., BERNHARD, R. A., MERSON, R. L. and PATINO, H. Evaluation of mechanical deaeration parameters for high vacuum canned foods. *J. Fd Sci.*, **53**(1988) 157–61.

54. ROBE, K. Vacuum-pack adds new dimension to flame sterilization. *Fd Proc.*, Sept. 1979, 102–4.

55. SCHWEINGRUBER, P. J., CARROAD, P. A., LEONARD, S. J., HEIL, J. R., WOLCOTT, T. K., O'MAHONY, M. and WILSON, A. Evaluation of instrumental methods for firmness measurements of fresh and canned clingstone peaches. *J. Text. Studies*, 1981, **12**, 389–99.

56. QUAST, D. G. and SIOZAWA, Y. Y. Heat transfer rates during heating axially rotated cans. *Proc. IV Int. Cong. Fd Sci. and Technol.*, 1974, **IV**, pp. 458–68.

57. SOULE, C. L. and MERSON, R. L. Heat transfer coefficients to Newtonian liquids in axially rotated cans. *J. Fd Engrg*, 1985, **8**, 33–46.

58. KAYS, W. M. and BJORKLUND, I. S. Heat transfer from a rotating cylinder with and without crossflow. *Tr. ASME Ser. C.*, 1958, **80**, 70–80.

59. LOHNING, F. M. Control instrumentation for flame sterilization. In *Flame*

Sterilization. Specialist Courses for the Food Industry, AIFST–CSIRO, North Ryde, NSW, Australia, No. 2, 1972, pp. 47–8.

60. DE RUYTER, P. W. and BRUNET, R. Estimation of process conditions for continuous sterilization of foods containing particulates. *Fd Technol.*, 1973, **25**(7), 44–60.

61. LEONARD, S., MARSH, G. L., MERSON, R. L., YORK, G. K., HEIL, J. R., FRYER, S., WOLCOTT, T. K. and ANSAR, A. Comparative procedures for calculating Steriflamme thermal processes. *J. Fd Sci.*, 1975, **40**, 250–3.

62. YORK, G. K., HEIL, J. R., MARSH, G. L., ANSAR, A., MERSON, R. L., WOLCOTT, T. K. and LEONARD, S. Thermobacteriology of canned whole peeled tomatoes. *J. Fd Sci.*, 1975, **40**, 764–9.

63. Biol. Resch. Lab. *Studies on flame-sterilization of canned fruit*, Report No. 579, Dec. 8, 1961.

64. YAWGER, E. S. The count reduction system of process lethality evaluation. In *Metodo Simplificado para la Determinacion del Poder Letal de un Proceso Termico*, Apartes de Informacion Conservera, 1967, **158**(7), Redaccida y administracion: Valencia, Spain.

65. NCA *Laboratory Manual for Food Canners and Processors, Vol. I, Microbiology and Processing*, AVI Publ. Co. Inc., Westport, CT, 1968.

66. Anon. Making sure canned food is sterilized. *CSIRO Ind. Res. News*, 1978, March, **127**.

67. CUMMING, D. B. and WRIGHT, H. T. A thermocouple system for detecting centre temperature in cans processed by direct flame sterilizers. *Can. Inst. Fd Sci. Technol. J.*, 1984, **17**(3), 152–6.

68. BEAUVAIS, M. 1978, private communication.

69. LEONARD, S., OSAKI, K. and HEIL, J. R. Monitoring flame sterilization processes. *Fd Technol.*, 1984, **38**, 47–50, 83.

70. GOLDONI, J. S., KOJIMA, S., LEONARD, S. and HEIL, J. R. Growing spores of P.A. 3679 in formulations of beef heart infusion broth. *J. Fd Sci.*, 1980, **45**, 467–70, 475.

71. TAMURA, M. S., SHOEMAKER, C. F. and HENDERSON, J. M. Real-time lethality evaluation during the flame sterilization of canned foods. *J. Fd Sci.*, 1985, **50**, 808–11.

72. CASIMIR, D. J. Economics of flame sterilization. In *Flame Sterilization. Specialist Courses for the Food Industry*, AIFST–CSIRO, North Ryde, NSW, Australia, No. 2, 1972, pp. 10–11.

73. FERRUA, J. P. and COL, M. H. Standardized tests compare energy consumption rates for sterilizing equipment. *Can./Pack.*, No. 1, 1975, 44.

74. SINGH, R. P. 1981, private communication.

Chapter 3

EFFECT OF THERMAL PROCESSING ON MILK

H. G. KESSLER

*Institute for Dairy Science and Food Process Engineering,
Technical University Munich, FRG*

SUMMARY

The purpose and the methods of heat treatment, especially the effect of temperature on milk bacteria, have been discussed. The destruction of microorganisms and chemical changes which result in damage to vitamins and amino acids, in browning and formation of hydroxymethylfurfural as a consequence of the Maillard reaction, and in whey protein denaturation have been investigated and described in terms of the laws of reaction kinetics. It has been shown that the experimental determination of the order and rate constants of reaction allow the calculation and prediction of changes with a relatively high degree of accuracy, not only for pasteurization, UHT-heating and sterilization but also for storage conditions. Milk constituents which are not affected by the dissolved oxygen in the milk, either during heating or during storage, have been considered. Chemical and sensory changes due to oxidative reactions have been excluded.

1. THE PURPOSE OF HEAT TREATMENT

The microorganisms contained in raw milk are completely or partially destroyed by heat treatment. In addition, enzymes are inactivated to an extent which depends on the temperature of heating and on its duration.

Heat treatments are chosen which preserve as far as possible the desirable organoleptic and nutritional qualities of milk and reduce the number of pathogenic organisms (pasteurization) or destroy all microorganisms and inactivate all enzymes (sterilization).

91

Pasteurization is carried out at temperatures below 100°C. As the activity of enzymes is only partially inhibited and the number of microorganisms only reduced, the product has a limited storage stability. Provided no recontamination occurs, pasteurized milk has a shelf life of 5–20 days, depending on the storage temperature. However, a shelf life of 20 days can only be achieved at a storage temperature below 5°C.

Thermization is usually carried out at temperatures ranging from 62–65°C and is aimed at the destruction of heat sensitive microorganisms.

During sterilization (carried out at temperatures above 100°C) all organisms which could grow during storage at room temperature are destroyed. Enzymes are also almost completely inactivated so that the product would keep for ever if organoleptic changes, caused by chemical reactions, would not limit its shelf life.

Most of the enzymes initially present in milk can be inactivated at temperatures below 100°C. But some enzymes derived from psychrophilic microorganisms such as pseudomonads need temperatures above 100°C and some proteinases and lipases need temperatures above 150°C for inactivation.

2. METHODS OF HEAT TREATMENT, THE EFFECT OF TEMPERATURE ON MILK BACTERIA

Milk contains acid-forming streptococci, bacilli, psychrophilic bacteria, micrococci, coli types and, in a few cases, spore formers. But milk from sick animals or handled by sick personnel may contain pathogenic organisms such as pathogenic coli types as well as organisms causing typhus, paratyphus, tuberculosis, brucellosis, salmonella and mastitis. Relatively mild treatment is sufficient to destroy pathogenic bacteria. This is shown in Fig. 1, for pasteurization. Some enzymes, e.g. alkaline phosphatase, are inactivated by heat treatment in the same time/temperature range.

2.1. Pasteurization

The name pasteurization is applied to a heat treatment whose time/temperature range is just a little above that necessary for the destruction of pathogenic microorganisms. The conditions of such heat treatment, the regulation for which may vary from country to country, are shown in Fig. 1.

—Batch heating   62–65°C for   30 min
—Short time heating   71–75°C for 15–40 s
—High temperature heating   85–90°C for   2–10 s

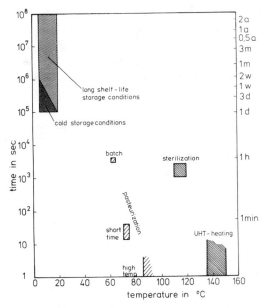

FIG. 1. Heating and storage conditions for milk.

The aim of pasteurization is to reduce the number of pathogens to such an extent that the consumption of the pasteurized milk does not constitute a health hazard.

Immediately after heat treatment pasteurized milk usually still contains heat resistant organisms initially present in raw milk, mainly *Streptococcus thermophilus* as well as micrococci and microbacteria. Spore formers, heat resistant lactobacilli and achromobacteria are present in lower numbers. To what extent the number of organisms can be reduced by pasteurization depends not only on the numbers present initially but also on the type of organism. Pasteurization kills heat sensitive microorganisms and therefore destroys pathogens. While short time pasteurization has little effect on heat resistant microorganisms, the high temperature heat treatment permitted in some countries may destroy a large proportion of the heat resistant flora, the precise extent depending on its composition. However, the reduction in the number of heat resistant organisms does not mean that the shelf life of milk is thereby prolonged. The organisms surviving pasteurization hardly multiply during the cold storage of milk.

Based on an extensive literature search and on her own investigations Niemierski[1] reports that the widely used short time heating at 72°C for 15 s (which are also the minimum heating conditions recommended by the International Dairy Federation (IDF)) is sufficient to destroy pathogens even *Salmonella senftenberg* and *Rickettsia burneti*, bacterial strains which are usually regarded as heat resistant. The same is true for *Listeria monocytogenes* in milk for the destruction of which a z-value of 6·3°C and a D-value at 71·1°C of 0·95 s was found.[2] This work led to the conclusion that *Listeria monocytogenes* does not survive pasteurization (the reduction in the number of organisms by 15 orders of magnitude by heating at 71·7°C for 15 s). It was also found that its heat resistance did not change.

2.2. Thermization

IDF document 97 (see van den Berg[3]) concludes from the scrutiny of a large number of publications that heating at 62–65°C for 15–20 s will kill heat sensitive microorganisms. It has been observed repeatedly that the number of psychotrophic bacteria, especially species of *Pseudomonas*, can increase to more than 10^5–10^6 per ml within 3 days of storage at 4°C even if the original contamination was small. The metabolic products of these bacteria, such as heat resistant lipases and proteinases impair the quality of milk. However, when milk is not recontaminated after thermization its bacterial count will remain low and its quality preserved, even after 3 days of storage at 7°C.

The mild heating conditions of thermization have no deleterious effect on the product. It must be pointed out, however, that thermization cannot replace pasteurization. It should be used where cold storage of the raw milk is unavoidable. Temperatures higher than 65°C and holding times longer than 20 s do not increase the effect.

2.3. Sterilization

The aim of sterilization, i.e. ultra high temperature (UHT) heating and heating in an autoclave is to destroy all microorganisms which could grow during normal storage at room temperature. It follows therefore that the whole of the microflora has to be destroyed, with the possible exception of some heat resistant spores such as those of *Bacillus stearothermophilus*, which could survive the heat treatment. However, these spores do not germinate during storage at room temperature. The conditions of UHT and autoclave heating are so designed that the destructive effect is the same in both cases although the temperature/time conditions of heating are different in the two methods.

—Sterilization (autoclave)   109–120°C for 40–20 min
—UHT heat treatment   135–150°C for 20–2 s

The two methods of heat treatment produce different chemical changes and they also differ in the extent to which heat resistant lipases and proteinases, produced by the psychrotrophic flora which may have multiplied too much in the raw milk, are inactivated. In cases where such enzymes determine the shelf life of the milk, heating in an autoclave is superior to UHT treatment. The quality of the raw milk should therefore always be good and it should be processed as quickly as possible.

3. THE REACTION KINETICS OF HEAT INDUCED CHANGES

The best process design for the heat treatment of foods is one in which the chosen temperature/time combination of heating makes certain that the desirable effects are obtained but deleterious changes are kept to a minimum.

Both heat treatment and storage conditions are defined mainly by temperature and time. By plotting the logarithm of time against the temperature all kinds of heat treatment and storage conditions can be depicted in one diagram, although the aims of these treatments are quite different from each other (Fig. 1). Thus, the main purpose of cold storage is the inhibition of the multiplication of those organisms which have survived pasteurization. Organoleptic spoilage correlates with a bacterial count that has increased, on average, to 2×10^6 per ml. Products which have been preserved by sterilization or UHT (ultra high temperature) treatment are not supposed to be subject to microbial spoilage. However, it is expected of them that chemical and organoleptic changes should remain small during prolonged storage. Thus the aim of the heat treatment is as described above.

3.1. Reaction Kinetics

To preserve the quality of the product as far as possible, the heat treatment should be the mildest possible. In order to be able to choose the best heat treatment and storage conditions for the preservation of the quality of the product it is necessary to know how heat-produced changes such as the destruction of microorganisms, the inactivation of enzymes, the denaturation of proteins, browning and the destruction of vitamins and amino acids are affected by time. However, it is only rarely possible to follow the real chemical molecular events.

It is therefore useful to deal with empirically obtained data on changes and to describe these in the formulation of kinetics. It is often possible to obtain insights into the real molecular reaction from these reaction kinetic treatments.

3.2. The Rate of Reaction

The rate of reaction is the change with time in concentration of the substance under investigation. For decomposition reactions it is $-dC/dt$ and for reactions leading to the formation of new compounds, $+dC/dt$. For decomposition reactions of the nth order of reaction the following equation is suitable:

$$-\frac{dC}{dt} = k_n \cdot C^n \tag{1}$$

The rate constant k_n of the reaction is influenced by temperature.

If the initial concentration of the substance C_o changes (by decomposition) to C_t after time t, then the ratio C_t/C_o can be calculated from eqn. (1).

$$\frac{C_t}{C_o} = [1 + (n-1)kt]^{1/(1-n)} \quad \text{(valid for } n \neq 1) \tag{2}$$

This equation describes a general forward reaction applying to any fractional or integral value of n except 1. In the special case of a reaction of first order ($n = 1$), C_t/C_o becomes

$$\frac{C_t}{C_o} = \exp(-kt) \tag{3}$$

In eqns (2) and (3) the rate constant k is

$$k = k_n \cdot C_o^{n-1} \tag{4}$$

Chemical changes may cause the formation of new substances F with the simultaneous decomposition of a substance R:

$$-\int_{C_{RO}}^{C_{Rt}} \frac{dC_R}{dt} = +\int_0^{C_{Ft}} \frac{dC_F}{dt} \qquad C_{RO} - C_{Rt} = C_{Ft} \tag{5}$$

In analogy to eqns (1) and (4) the reaction or the formation of substances may be described by eqns. (6) and (7):

$$\text{when } n \neq 1: \ 1 - \frac{C_{Ft}}{C_{RO}} = [1 + (n-1)kt]^{1/(1-n)} \tag{6}$$

$$\text{when } n = 1: \ 1 - \frac{C_{Ft}}{C_{RO}} = \exp(-kt) \tag{7}$$

In these reactions C_{Ft} is often very much smaller than $C_{RO}(C_{Ft} \ll C_{RO})$. It can therefore be described with a good approximation to correctness as a reaction of zero order:

$$C_{Ft} = k_F \cdot t \qquad (8)$$

The order of the reaction can be determined with sufficient accuracy only when the changes in the concentration of the substances are fairly large.

The temperature dependence of the rate constants k or k_F can be expressed by the Arrhenius equation;

$$k = k_O\left(\frac{1}{T} = 0\right) \exp\left(-\frac{E_a}{R} \cdot \frac{1}{T}\right) \qquad (9)$$

where E_a is the energy of activation in J/mol, R is the universal gas constant ($= 8 \cdot 314 \, J/(mol \times K)$), and T is the absolute temperature in K.

3.3. Description of Identical Effects by Means of a Semi-logarithmic Time/Temperature Plot

Whatever the order of the reaction the following relationship holds for the temperature/time combinations which produce the same degree of decomposition or formation of new compounds.

$$k_1 \cdot t_1 = k_2 \cdot t_2$$

or

$$t_1 \cdot k_O \exp\left[-\frac{E_a}{RT_1}\right] = t_2 \cdot k_O \exp\left[-\frac{E_a}{RT_2}\right]$$

or

$$\ln\frac{t_1}{t_2} = -\frac{E_a}{R}\left[\frac{1}{T_2} - \frac{1}{T_1}\right] \qquad (10)$$

If a reference point with a temperature/time combination (T_R, t_R) is chosen for the description of identical effects then:

$$\ln\frac{t}{t_R} = -\frac{E_a}{R}\left[\frac{1}{T_R} - \frac{1}{T}\right] \qquad (11)$$

When by the use of this equation the logarithm of the time is plotted against the reciprocal of the absolute temperature (Fig. 2), identical effects are shown to lie on a straight line which passes through the reference value (t_R, T_R) and whose slope is $(-E_a/R)$ or $(-E_a/2 \cdot 303\,R)$ when the logarithm to the base 10 has been used.

Depending on the order of the reaction, it is possible to calculate other

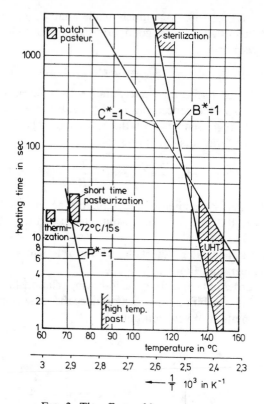

FIG. 2. The effects of heat treatment.

effects (C/C_o = constant) point by point and to draw parallel straight lines because all the slopes are the same.

The relevant equation, valid for reactions of the nth order when $n \neq 1$ is:

$$\frac{C_t}{C_o} = \left[1 + (n-1)k_o \cdot t \cdot \exp\left(-\frac{E_a}{RT} \right) \right]^{1/(1-n)} \tag{12}$$

The equation valid for a reaction of the 1st order when $n = 1$ is:

$$\frac{C_t}{C_o} = \exp\left[-k_o \cdot t \cdot \exp\left(-\frac{E_a}{RT} \right) \right] \tag{13}$$

Sometimes a plot of the logarithm to the base 10 of time against the temperature in °C [$\log t = f(\vartheta \text{ in } °C)$] is encountered. The following relationship then applies:

$$\log \frac{t}{t_R} = -\frac{E_a}{2 \cdot 303 \, R T_R T}(\vartheta - \vartheta_R)$$

$$= -\frac{1}{z}(\vartheta - \vartheta_R) \qquad (14)$$

where

$$z = \frac{2 \cdot 303 \, R T_R T}{E_a} \qquad (15)$$

For narrow temperature ranges in which the temperature dependence of the z-value can be neglected, the z-value expresses the increase in temperature in °C or in K which would produce the same effect in one-tenth of the time. However, in reality the relationship is not linear and this must be taken into account when wider temperature ranges are considered. The results of many investigations have shown that the activation energy of a change is constant and independent of the temperature. This was observed for a large number of reactions.

4. BACTERIOLOGICAL EFFECTS

The destruction of microorganisms is generally well described as a reaction of the first order, i.e. the logarithm of the bacterial count decreases linearly with time. The following equation makes it possible to establish the destructive effects of a particular heat treatment including its warming up and cooling down phases:

$$\log \frac{C_o}{C} = \frac{1}{D_{\vartheta_R}} \int_0^t 10^{\frac{\vartheta - \vartheta_R}{z}} \, dt \qquad (16)$$

$\log(C_o/C)$ denotes the number of the orders of magnitude by which the bacterial count is reduced (initial count, C_o; count at time t, C). D_{ϑ_R} is the decimal reduction time at the reference temperature ϑ_R.

The product $[D_{\vartheta_R} \cdot \log(C_o/C)]$ is the so-called F-value. In the canning industry a temperature of 121°C ($\approx 250°F$) is often taken as the reference temperature and the so-called F_o-value is defined by the following equation, and refers to 12 times the decimal reduction time ($12 \, D_{121°C}$) for *Clostridium botulinum* at $z = 10°C$:

$$F_o\text{-value} = D_{121°C} \cdot \log \frac{C_o}{C} = D_{121°C} \cdot 12$$

The F_o-value, with its reference temperature of 121°C, might be suitable for the sterilization of canned goods because these are sterilized in the relatively narrow range of $121°C + 5°C$. However, if the temperature range is to include pasteurization and UHT treatment then the effect of temperature on the z-value has to be taken into account. The temperature at which microorganisms are destroyed depends on the genus. Heat sensitive microorganisms are killed at temperatures between 55 and 70°C, heat resistant ones between 70 and 90°C, mesophilic spores between 110 and 130°C and thermophilic spores between 110 and 150°C. Both mixed cultures and pure cultures have been examined and z-values were found ranging from 9–11·5 K for the spores of *B. stearothermophilus*, *B. subtilis* and *B. coagulans*. The lower values were generally found at the lower temperatures. A large number of investigations with heat sensitive microorganisms (Gram negative, psychrophilic) gave z-values from 6–9 K at temperatures between 68 and 74°C.

These results demonstrate the temperature dependence of the z-values in the temperature range appropriate for the destruction of any particular microbial species. It is therefore appropriate to reckon with a constant activation energy also for the destruction of microorganisms. The value of the activation energy compares with that found for the unfolding of the structure of the whey proteins. A value of $E_a = 285\,000$ J/mol is suggested. When this value is kept constant eqn (15) yields the following temperature/ z-value combinations.

Temperature:	(°C)	65	75	100	125	150
z-value:	(K)	7·6	8·1	9·3	10·6	12·2

4.1. The Effects of Pasteurization

To compare pasteurization treatments the recommendations of the IDF for minimum conditions for pasteurization, namely 72°C for 15 s, are used as a basis and if these conditions are assigned a dimensionless number for the pasteurization effect namely $P^* = 1$, then other temperature/time combinations producing the same effect can be calculated by eqn. (17):

$$\ln\frac{t}{t_o} = \frac{E_a}{R}\left(\frac{1}{T} - \frac{1}{T_o}\right) \tag{17}$$

where $E_a = 285\,000$ J/mol (energy of activation when $z = 8$ K
 and $\vartheta = 72°C$)
 $R = 8\cdot314$ J/mol K (universal gas constant)
 $t_o = 15$ s (relevant holding time)
 $T_o = 345$ K (absolute temperature of the relevant
 temperature $\vartheta = 72°C$)

The value of $E_a = 285\,000$ J/mol was chosen as the activation energy for the destruction of microorganisms. This value was derived from a z-value of 8 K which can be regarded as a good average value for the destruction of many strains of microorganisms in the temperature range of pasteurization (71–75°C).

The only time/temperature combinations which give the same pasteurization effect $P^* = 1$ can be determined from the following equation which is obtained by substituting the appropriate values in eqn. (17):

$$\log t = \frac{14\,885}{T} - 41\cdot97 \tag{18}$$

(t in s; T in K)

This relationship is only valid in the temperature range between 71 and 80°C.

Figure 2 shows the straight line $P^* = 1$ which passes through the point 72°C/15 s. Figure 3 depicts the pasteurization effects P^* derived from $P^* = 1$. Proper pasteurization has therefore only been achieved when $P^* > 1$. If, for example, pasteurizing conditions which give $P^* = 1$ reduce the number of pathogenic organisms by 10^{12}, then conditions of $P^* = 4$ would reduce the numbers by 10^{48}. Equation (19) yields the pasteurization effect for a given heat treatment (including warming up, holding, and cooling down).

$$P^* = \int_0^t \frac{dt}{t_{P^*=1}} \tag{19}$$

$$\frac{1}{t_{P^*=1}} = 10^{41\cdot97 - (14885/T)} \tag{20}$$

(t in s; T in K)

This equation can be solved either numerically or graphically. The latter solution is shown in Fig. 4. Pasteurization effects for temperature/time

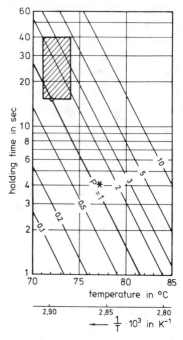

FIG. 3. Pasteurization effects based on the IDF recommendation of 72°C 15 s.

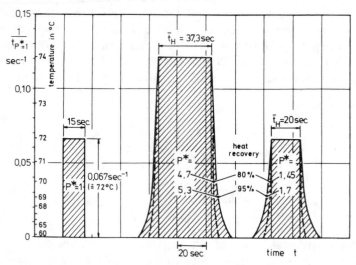

FIG. 4. Evaluation of pasteurization effects of heating processes.

treatments of 74°C with holding times ranging from 30–40 s and 72°C with holding time of 20 s are shown. When $1/t_{P^*=1}$ is plotted against the time (its value having been obtained from eqn. (20) or Fig. 3 then the shaded area represents the integral of eqn (19) and therefore the value found for P^*. The size of the area for $P^* = 1$ (15 s/72°C) is shown for comparison. In addition, the effects of warming up and cooling down in plate heat exchangers with 80% or 95% heat recovery are depicted. They are seen to be relatively small. The shape of the graph shows that a pasteurization effect of $P^* = 5.3$ is excessive and could be reduced by lowering the temperature and the holding time.

4.2. The Effects of UHT Heating and Sterilization

Using eqn (17) the effects can be presented in a plot of log t against $1/T$ as straight lines. The minimum heat treatment necessary for obtaining a milk of good keeping quality was laid down as being that killing rate of thermophilic spores (from the mixed flors of a milk as delivered to the dairy) which reduced the spore count by 10^9. The z-value of 10.4 K obtained by Horak[4] at 121°C yields the same value for the energy of activation namely $E_a = 285\,000$ J/mol, as the z-value of 8 K in pasteurization. If it is assumed, as seems justified, that the energy of activation of the destruction of most bacteria is nearly the same, then the z-value at 145°C would be 11.7 K.

To denote the minimum heating effect the dimensionless term $B^* = 1$, the so-called bacteriological effect was introduced.[5,54] It represents a reduction in the thermophilic spore count by 10^9. The various temperature/time combinations which give a value of $B^* = 1$ (see also Fig. 2) always produce the same bacteriological effect. When $B^* = 2$ the reduction in the number of spores is by definition $9 \times 2 = 18$ powers of ten.

The minimum temperature/time combination for $B^* = 1$ is given by the following equation:

$$\log t = \frac{14\,885}{T} - 35.51 \qquad (21)$$

(t in s; T in K)

5. CHEMICAL CHANGES

From a nutritional point of view it is important to know whether essential amino acids and vitamins are lost. It is known that lysine suffers due to the Maillard reaction. A large number of investigations of the effects of pasteurization have failed to show any loss of the biological quality of the proteins. The higher temperatures used in sterilization do, however, produce small losses of arginine, lysine, histidine, methionine, leucine, valine and cystine. UHT heating however, hardly impairs the nutritional value of the protein because the non-enzymatic browning reaction does not take place.

Vitamin losses due to heating are very small on milk heat treatment, i.e. pasteurization and UHT treatment. Losses are little greater for milk sterilized in cans or bottles. The type of heating, i.e. whether direct or indirect, is important for vitamins B_6, B_{12} and folic acid. Indirect heating produces almost twice the loss of vitamin B_6 and folic acid than direct heating with steam does, while for vitamin B_{12} the situation is reversed.[6]

It should again be pointed out that fat has a protective effect. It considerably reduces the heat-induced vitamin loss, particularly in the case of the fat-soluble vitamins A, D, E and K. Certain water-soluble vitamins such as riboflavin (B_2), nicotinic acid (B_3), pantothenic acid (B_5) and biotin (B_8) suffer hardly any losses, while thiamin (B_1), pyridoxine (B_6), cyanocobalamine (B_{12}) and ascorbic acid (C) are considerably more sensitive to heat.[7,8] The milk should be degassed (removal of oxygen) as completely as possible to preserve in particular vitamin C and folic acid.

In order to choose the best heat treatment methods and to achieve the best possible product quality it is necessary to establish experimentally the effect of the time and temperature of the heat treatment on the decomposition and formation reactions and to describe the changes, if possible, in terms of reaction kinetics.

One must be aware, however, that the experimentally derived order of the reaction is not necessarily the same as the real order of the reaction between the molecules. Changes in a compound can take place by means of intermediate steps in which several substances participate. However, the experimental determination of the order of reaction affords the only possibility of calculating and predicting changes with a relatively high degree of accuracy and of designing processes accordingly.

In the following we are going to consider milk constituents which are not affected by the dissolved oxygen in the milk either during heating or during storage. Chemical and sensory changes due to oxidative reactions, such as the influence on free SH-groups, ascorbic acid, on flavour and odour are excluded.

Horak,[4] Fink[9] and Dannenberg[10] have studied the effects of heat treatments over a wide temperature range on the reduction in the concentration of heat sensitive, nutritionally important milk constituents such as vitamin B_1 (thiamin), on the loss of the essential amino acid lysine and the formation of the reaction products of the Maillard reaction (hydroxymethylfurfural (HMF) and browning products). The investigations were carried out in the temperature ranges used for storage, pasteurizing up to sterilization and UHT treatment and periods of time were as long as possible. In addition, the rates of heat induced denaturation of whey proteins were studied.

Reports in the literature on the losses of thiamin during UHT treatment of milk differ considerably. Kon,[11] Nagasawa et al.,[12] Hostettler,[13] Gregory and Burton,[14] Ford et al.,[15] Karlin,[16] Burton et al.,[17] and Horak[4] state the values for thiamin losses to be between a minimum of 0% and a maximum of 3%. Others, however, such as Lembke et al.,[18] Görner and Uherová[19] and Bayoumi[20] on the other hand, found losses of up to 20%. The results obtained for stored UHT milk are quite similar. Depending on the oxygen concentrations after processing (8·4 up to <1·0 mg/litre), Thomas et al.[21] did not detect any loss of thiamin during storage (23°C, 62 days) in any of the samples.

Reports in the literature of losses of available lysine during UHT heating and storage of milk also vary considerably. Finot et al.,[22] determined 0%; Horak,[4] <1%; Blanc et al.,[23] 0·4–0·8% and Mottar and Naudts,[24] 4·3–6·5% after UHT heating. Töter[25] discovered no deterioration after a period of 6 months of storage at 4°C and 20°C but losses of 14–28% at 38°C.

The influence of UHT heating and storage of milk on the formation of non-enzymatic browning products, such as HMF or colour, were examined, among others, by Mottar and Naudts,[24] Renner and Dorguth,[26] Konietzko,[27] Horak,[4] Zadow,[28] Mogensen and Poulsen[29] and Bosset et al.[30] The results obtained are not always a source of valuable information, since it is not always possible to establish whether the HMF values reported represent the difference between those in raw and in processed milk.

5.1. Thiamin
The initial concentration of thiamin in milk ranged from 0·35–0·45 mg/litre. Temperature/time dependent results of the measurements[31,32] were inserted into eqns (2) and (3)—taking values of n of 0, 1, 2 and 3—until the best linear relationship was found.

Figure 5 depicts the loss of thiamin with time for various temperature/time combinations of heating and storage in the plot $(C_t/C_o)^{-1}$ versus time.

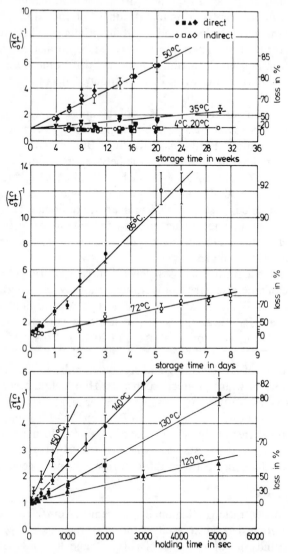

FIG. 5. Loss of thiamin in milk, plotted according to a 2nd order reaction (results from 120–150°C by Horak[4] and from 4°C to 85°C by Fink[9]). Reproduced from Kessler and Fink,[31] © 1986 Institute of Food Technologists.

TABLE 1

RATE CONSTANTS k AND COEFFICIENTS OF CORRELATION r FOR THIAMIN LOSSES IN MILK AS A FUNCTION OF TEMPERATURE

Temperature ($°C$)	n	r	k (s^{-1})
35	52	0·8811	$6·60 \cdot 10^{-8}$
50	41	0·9926	$4·14 \cdot 10^{-7}$
72	10	0·9931	$4·49 \cdot 10^{-6}$
85	10	0·9958	$2·28 \cdot 10^{-5}$
120	7	0·9955	$3·01 \cdot 10^{-4}$
130	8	0·9971	$8·29 \cdot 10^{-4}$
140	8	0·9993	$1·52 \cdot 10^{-3}$
150	6	0·9980	$2·84 \cdot 10^{-3}$

Reproduced from Kessler and Fink,[31] © 1986 Institute of Food Technologists.

The range of the deviation of the measured values is indicated. It can be seen that all values obtained at the same temperature lie on a straight line. The least square fits were expressed by the coefficient of correlation r, given in Table 1. It follows from eqn (2) that this is only possible if $n = 2$. The decomposition of thiamin can be described as a second order reaction:

$$\left(\frac{C_t}{C_o}\right)^{-1} = 1 + kt \qquad (22)$$

It is furthermore evident from the values of thiamin losses given as $\% = 100 (1 - C_t/C_o)$ on the right hand side of the ordinate in Fig. 5, that the order of the reaction can only be determined with a satisfactory degree of reliability if the losses are more than 50%.

The rate constants k (Table 1) follow from the slopes of the straight lines using eqn (22). These are shown in the Arrhenius plot of Fig. 6, where $\ln k/k_o$ is plotted against the reciprocal of the absolute temperature. $k_o(1/T = 0)$ is found by means of the k-values at different temperatures and using eqn (9). The result is surprising. It could not be assumed a priori that the Arrhenius equation would be valid over such a large temperature and even larger time range, especially as the actual nature of the reaction on the molecular level is unknown.

The value of the energy of activation can now be found from the slope of the straight line in Fig. 6, which is obtained by the least square fit, taking into consideration the total population of data ($r = -0·9994$). The temper-

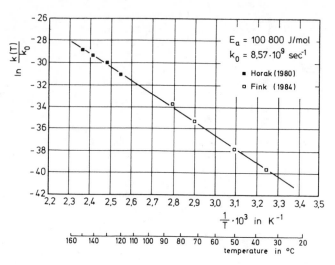

FIG. 6. Arrhenius plot for the loss of thiamin in milk. Reproduced from Kessler and Fink,[31] © 1986 Institute of Food Technologists.

ature/time dependent losses L are given by the following equation, derived from eqns (2) and (9).

$$L = 1 - \frac{C_t}{C_O} = 1 - \left[1 + k_o \cdot t \cdot \exp\left(- \frac{E_a}{RT} \right) \right]^{-1} \qquad (23)$$

where $k_o = 8.57 \times 10^9$ s^{-1}; $R = 8.314$ J/mol K; $E_a = 100\ 800$ J/mol.

Equation (23) is used to obtain the data shown in Fig. 7, which is a plot of the logarithm of time against the reciprocal of the absolute temperature and shows a number of straight lines of equal thiamin losses. These data can be compared with experimentally obtained ones, shown as dots in the graph. It is therefore easily possible to predict the effect of the most varied heat treatment conditions on the changes in thiamin content. It is even possible to obtain a quantitative value for the effects of storage. Thus sterilization in the autoclave produces the same losses as storage at 20°C for 1–2 years, and UHT treatment, e.g. 140°C for 22 s, has the same effect as storage of 6 weeks at 20°C. Although it is not possible to measure the losses in thiamin concentration in pasteurized milk, calculation indicates that these range from 0·1–0·01%; the extrapolation of calculated values shows that pasteurization produces the same effect as storage for 1–7 days at 5–10°C.

A criterion for the heat stress was the C^* value, introduced and defined by

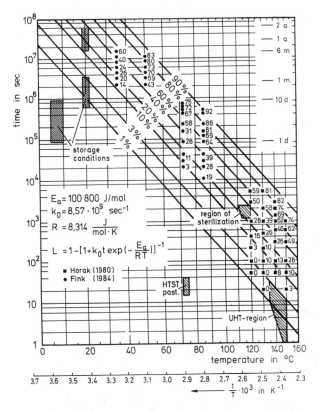

FIG. 7. Loss of thiamin in milk, calculated and measured (values at measuring points stated in percent of loss). Reproduced from Kessler and Fink,[31] © 1986 Institute of Food Technologists.

Kessler.[5] C^* is a chemical effect based on the loss of thiamin. $C^* = 1$ means a thiamin loss of 3%. If the loss in thiamin is not more than given by the limiting line (Fig. 2) for a 3% loss, calculated from measurements of higher thiamin losses, very little damage will be caused to the product. It must be borne in mind that no thiamin losses were detected in the range of this 3% line by the analytical method used.

5.2. Lysine

Experiments[31,33] using a wide range of temperatures and times showed that the losses of lysine (namely the reduction in the nutritional availability of this essential amino acid) could also be described by a second order reaction, i.e. by eqn (23). From the slope of the straight line of the Arrhenius plot the following reaction kinetic constants were found for lysine:

$$E_a = 109\ 000\ \text{J/mol}$$

$$k_0 = 1{\cdot}95 \times 10^{10}\,\text{s}^{-1}$$

Figure 8 in which lysine losses calculated from eqn (23) are compared with data obtained experimentally, shows that this equation is valid over a wide temperature/time range.

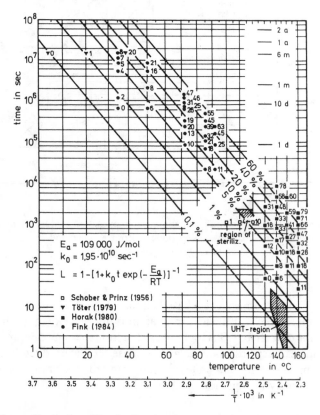

FIG. 8. Loss of lysine in milk, calculated and measured (values at measuring points stated in percent of loss). Reproduced from Kessler and Fink,[31] © 1986 Institute of Food Technologists.

5.3. Colour Changes

The end products of the Maillard reaction are brown compounds, the melanoids. The presence of these compounds indicates a more or less severe heat treatment. Horak[4] and Fink[9] investigated the changes taking place in whole milk after special heat treatments and storage conditions. A filter colorimeter was used for the colour measurements which permitted a precise characterization of each colour by two colour parameters and the brightness.

Figure 9 shows colour changes in heated and stored whole milk. The graph shows that equal changes in colour can be represented as straight

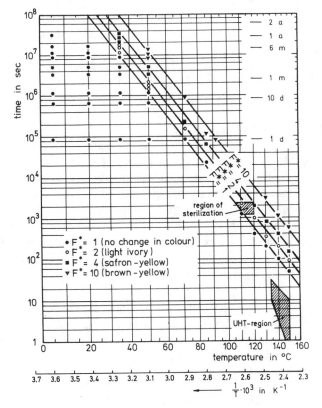

FIG. 9. Lines of identical colour changes for milk.[5,9] Reproduced from Kessler and Fink,[31] © 1986 Institute of Food Technologists.

lines when the logarithm of the time is plotted against the reciprocal of the absolute temperature. It is again evident that colour changes can be predicted from an Arrhenius plot. Equal amounts of heat stress lead to the same colour changes and this is independent of the temperature.

The activation energy E_a can be calculated from the slope of the straight lines in Fig. 9 and from the knowledge that the product of the rate constant k and the reaction time i is constant ($k_1 \cdot t_1 = k_2 \cdot t_2$ for temperatures 1 and 2):

$$\log \frac{k_r}{k_{r,1}} = \log \frac{t_{r,1}}{t_r} = -\frac{E_a}{2 \cdot 303\, R}\left(\frac{1}{T_r} - \frac{1}{T_{r,1}}\right) \tag{24}$$

E_a was found to be 116 kJ/mol. The suffix r,1 stands for a reference temperature $T_{r,1}$ (chosen at will) and the corresponding time $t_{r,1}$. T_r and t_r are the variable temperature/time combinations which lie on the reference line $\log t_r = f(1/T_r)$, i.e. on the straight lines of the equal colour. The position of the straight line is given by the point $(T_{r,1}; t_{r,1})$, and its slope by the energy of activation.

In Fig. 9, F^*-values represent a degree of colour change. $F^* \leqslant 1$ signifies no colour change. To make the F^*-values somewhat more illustrative, Fig. 9 shows some of the degrees of colour changes with the respective colours.

5.4. Hydroxymethylfurfural (HMF)

HMF is an intermediate product in the formation of brown pigments during the Maillard reaction which takes place on heating milk. The formation of HMF and the kinetics of this reaction were studied using the same wide temperature/time range mentioned above. Because the amount of HMF formed is extremely small compared to the amount of the compounds from which it is produced, it is possible to describe HMF formation as a reaction of zero order (eqn (25)). This is confirmed by Fig. 10.

$$C_{HMF}(\mu mol/litre) = k_{OF} \cdot t \cdot \exp\left(-\frac{E_a}{RT}\right) \tag{25}$$

where k_{OF} is the rate constant for $(1/T) = 0$ and formation.

The slopes of the straight lines in Fig. 10 give the rise constants which were used to construct the Arrhenius plot shown in Fig. 11. The reaction kinetic data obtained from this graph were:

$$k_{OF} = 1 \cdot 73 \times 10^{17} \frac{\mu mol}{litre\, s}$$

$$E_a = 139\,000\, J/mol$$

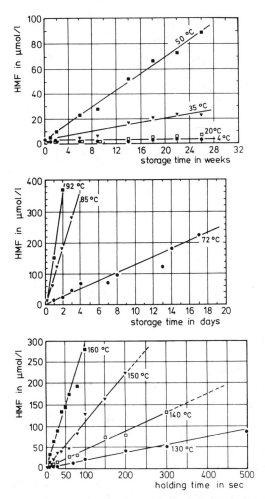

FIG. 10. Formation of HMF in milk, plotted according to a zero order reaction. Reproduced from Kessler and Fink,[31] © 1986 Institute of Food Technologists.

Equation (25) is again valid over a wide temperature range. Values for HMF contents, calculated from the equation, were plotted in Fig. 12 which shows that there is a linear relationship between the logarithm of time and the reciprocal of the absolute temperature.

It is clear from Fig. 12 that a UHT treated milk (135–150°C; 1–30 s) with

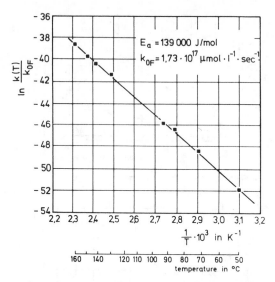

FIG. 11. Arrhenius plot for the HMF (hydroxymethylfurfural) formation in milk. Reproduced from Kessler and Fink,[31] © 1986 Institute of Food Technologists.

HMF values ranging from 1–10 µmol/litre is distinctly different from a sterilized (autoclaved) one (112–120°C; 1200–2400 s) containing 40–130 µmol/litre of HMF. Even if the method of HMF determination of Keeney and Bassette[34] registers newly formed substances of as yet undetermined identity, the present method makes it possible to assess the amount of thermal stress that a particular heat treatment delivers to the milk.

There is good agreement in the literature about the 'blank HMF value' that has to be subtracted from the measured ones. The research by Fink[9] has produced a relatively constant HMF value of 4·8 µmol/litre ($n = 109$; standard deviation (s) = 0·6 µmol/litre) with a range of 3·6–7·3 µmol/litre. Keeney and Bassette[34] (1959) give a value of 5 µmol/litre of raw milk and Konietzko[27] a range of 2·7–5·3 µmol/litre. The latter author has also found HMF values of 8·7 µmol/litre. A difference in the HMF values between raw and pasteurized milk could not be found.

5.5. Whey Protein Denaturation
Heat-induced changes in whey proteins in model solutions as well as in milk and whey have each received considerable attention.[35] Whey protein

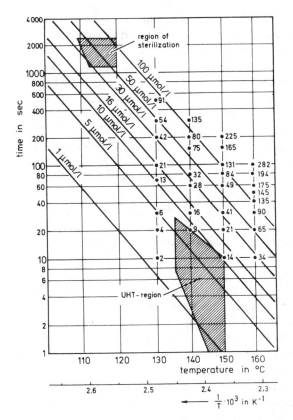

FIG. 12. HMF formation in heated milk (values at measuring points stated in percent of loss). Reproduced from Kessler and Fink,[31] © 1986 Institute of Food Technologists.

denaturation can produce a number of undesired effects such as deposits, but it can also be used to obtain desired modifications and improvements in products (e.g. texture of yoghurt) or processes (e.g. yield in production of fresh cheese or running time of heating plants by reduction of deposit formation).[55] One basic requirement in setting up operating conditions to produce a desired effect is a knowledge of the denaturation behaviour of individual whey proteins. A study of the kinetics of the reactions of denaturation should be helpful. There are a number of kinetic studies of the denaturation of isolated whey proteins in buffer solutions.[36–43,53]

Only a few studies using milk concern the behaviour of all of the whey protein fractions taken together.[44,45] The denaturation of individual whey protein fractions (β-LG A and B, α-LA, BSA) was studied in unchanged milk and whey systems by Larson and Rolleri[46] (however, without reaction kinetic investigations), by Lyster[47] and by Hillier and Lyster.[48] The last two workers established activation parameters over a wide temperature range and they first pointed out that there was a break in the Arrhenius plot.

The 'degree of denaturation values' ($1-C_t/C_o$) discussed below refer to the denaturation of whey proteins. Denatured whey proteins are defined as those which, after precipitation by acid, do not appear any more on those places on an electrophoresis gel on which the native proteins have previously been located by isoelectric focussing.

Dannenberg[10] studied the denaturation of the main whey protein fractions, namely β-lactoglobulin (LG) A and B and α^{1-1}-lactalbumin (LA). He developed a quantitative method by adapting the highly efficient method of ultrathin isoelectric focussing to the determination of whey protein fractions.

Figure 13 shows that the denaturation of β-LG B (in skim milk) can be described as a reaction of the order $n = 1.5$. The same was found for the genetic variant β-LG A. However, the denaturation of α-LA was clearly a reaction of the first order ($n = 1$) as shown in Fig. 14.

It should be pointed out that the accurate determination of the order of reaction e.g. $n = 1.5$ for β-LG was only possible because degrees of denaturation of up to 90% were used in the calculation. When the degree of denaturation was below 50% it was impossible to distinguish between first, second or 1.5 orders of reaction.

The rate constants calculated from the straight lines obtained by regression analysis in Figs 13 and 14 were plotted against the reciprocal of the absolute temperature according to the Arrhenius equation [$k = k_o \exp(-E_a/RT)$]. Figure 15 shows the results for the three protein fractions studied. It is evident that the relationship ($\ln k$ versus $1/T$) is strictly linear in a given temperature range which makes it possible to determine the activation energy E_a. The graph shows that for each protein fraction, there are two temperature ranges in which the activation energies are markedly different. This manifests itself by differences in the slopes of the straight lines in the two temperature ranges. The break in the straight line is about 90°C for the two genetic variants of β-LG but at about 80°C for α-LA. Table 2 gives the activation energies (obtained by a least squares fit), the logarithm of the pre-exponential factor k_o (obtained from the Arrhenius equation), and the coefficient of correlation r^2 of the straight lines.

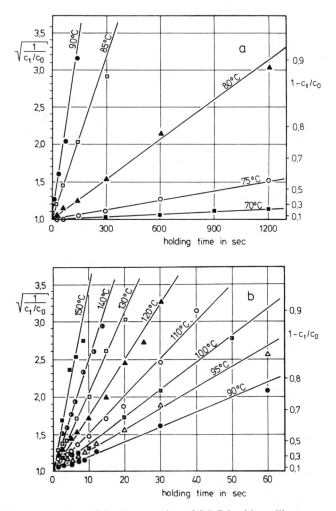

FIG. 13. Representation of the denaturation of β-LG in skim milk as a reaction of the order 1·5. Reproduced from Dannenberg and Kessler,[52] © 1988 Institute of Food Technologists.

From eqn (12) for β-LG and eqn (13) for α-LA as well as the reaction kinetic data found by experiment, the degrees of denaturation can be calculated. These are shown, together with measured values in Fig. 16 (β-LG) and in Fig. 17 (α-LA) as lines of equal degrees of denaturation. To make their position clearer the known milk heat treatment conditions have

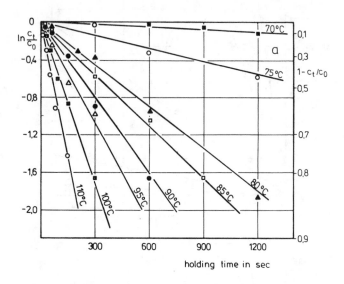

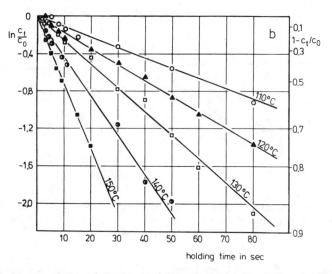

FIG. 14. Representation of the denaturation of α-LA in skim milk as a reaction of the first order. Reproduced from Dannenberg and Kessler,[52] © 1988 Institute of Food Technologists.

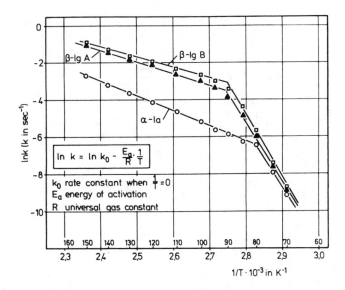

FIG. 15. Effect of temperature on the rate constant k for the denaturation of β-LG A and B, and α-LA in skim milk (k and k_o in s^{-1}). Reproduced from Dannenberg and Kessler,[52] © 1988 Institute of Food Technologists.

TABLE 2

REACTION KINETIC DATA FOR THE DENATURATION OF β-LG A AND B AND α-LA

Protein	n	Temperature (°C)	E_a (kJ/mol)	$\ln(k_o)$	r^2
β-LG A	1·5	70– 90	265·21	84·16	0·996
		95–150	54·07	14·41	0·997
β-LG B	1·5	70– 90	279·96	89·43	0·995
		95–150	47·75	12·66	0·999
α-LA	1·0	70– 80	268·56	84·92	0·997
		85–150	69·01	16·95	0·999

r^2: coefficient of correlation of the straight lines in Fig. 15.
Reproduced from Dannenberg and Kessler,[52] © 1988 Institute of Food Technologists.

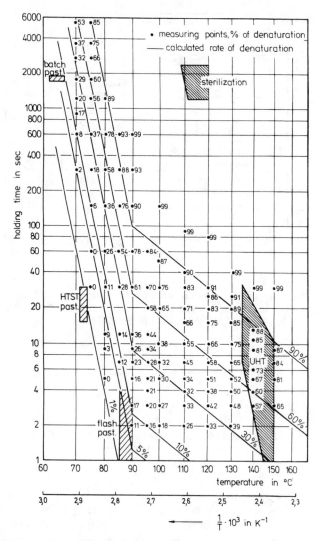

FIG. 16. Denaturation of β-lactoglobulin B in milk. Reproduced from Dannenberg and Kessler,[52] © 1988 Institute of Food Technologists.

been added to the diagram. The (previously observed) breaks in the straight lines of the Arrhenius plot (Fig. 15) are also evident in these diagrams. This discontinuity makes it clear that results obtained at low temperatures must not be extrapolated to higher ones and vice versa.

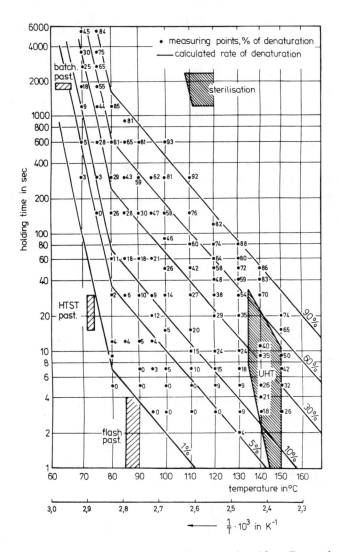

FIG. 17. Denaturation of α-lactalbumin in milk. Reproduced from Dannenberg and Kessler,[52] © 1988 Institute of Food Technologists.

The only explanation of this unusual effect of temperature on the rate constant is that there are two different reactions taking place, each of which dominates in a particular temperature range. In order to study these reactions Dannenberg[10] determined first of all the free activation enthalpy

ΔG^* using eqn (26) which is well known in chemical thermodynamics:

$$k = \frac{k_B \cdot T}{h} \exp\left[\frac{-\Delta G^*}{RT}\right]$$ (26)

He then used the relationship

$$\Delta H^* = E_a - RT$$ (27)

to establish the corresponding activation enthalpies ΔH^* where:

$k_B = 1 \cdot 381 \cdot 10^{-23} \, \text{J K}^{-1}$ Boltzmann constant
$h = 6 \cdot 624 \cdot 10^{-34} \, \text{J s}$ Planck constant
represents the index of the activation process

The changes in the free activation enthalpy are expressed as follows:

$$\Delta G^* = -RT \ln K^*$$ (28)

$$\Delta G^* = \Delta H^* - T\Delta S^*$$ (29)

K^* = the equilibrium constant of the formation of the activated complex
S^* = entropy of activation

Using eqn (26) k becomes:

$$k = \frac{k_B \cdot T}{h} \exp\left[\frac{\Delta S^*}{R}\right] \exp\left[\frac{-\Delta H^*}{RT}\right]$$ (30)

from which, knowing the value of ΔH^*, the entropy of activation ΔS^* can be found:

Protein	Temperature (°C)	$\Delta S^*(kJ/mol/K)$
β-LG A	75– 90	0·445
	95–150	−0·136
β-LG B	70– 90	0·487
	95–150	−0·150
α-LA	70– 80	0·452
	85–150	−0·115

The decrease in the value of the entropy in the high temperature range are remarkable. This suggests that the activated complex has a more ordered structure than the original compound.

To make it easier to explain the reasons for the break in the Arrhenius plot of the denaturation of whey proteins Fig. 16 is simplified and presented in Fig. 18 for β-LG B. This depicts, as examples, 3 lines for 3 different degrees of denaturation of β-LG B. It is evident that the lines all show

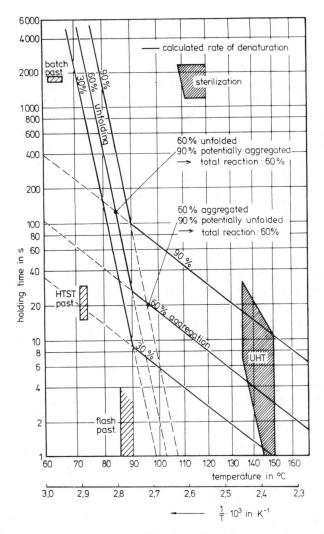

FIG. 18. Lines of equal degrees of denaturation drawn to demonstrate the temperature dependent changes in the rate constant.

a break at 90°C. The lines of equal degrees of denaturation represent temperature/time combinations in which both reactions, namely unfolding ones and aggregation ones, have taken place.

It is assumed that the unfolding reaction, represented by the steep lines, is rate determining for the whole denaturation process in the temperature range below 90°C. The example in Fig. 18 at 85°C shows that 60% of the β-LG B is unfolded and aggregated. Although 90% of the protein could be aggregated at this temperature (as deduced from the course of the aggregation reaction depicted by the flat part of the line), this is not possible because only unfolded proteins can aggregate and the unfolding reaction permits only 60% of the protein to be unfolded in this stage. The second example (at about 95°C) shows the rate determining effect of the aggregation reaction. Although 90% of the β-LG B is present in the unfolded form the aggregation reaction permits a degree of denaturation of only 60%.

This interpretation of the data is supported by the measured change in entropy ΔS^*. While the rate determining or rate limiting unfolding reaction (which is reversible in the absence of aggregation) produces an increase in entropy ΔS^* which indicates a state of greater disorder (unfolding), the negative ΔS^* values at the higher temperatures suggest that a state of greater order has been produced (aggregation).

Dannenberg,[10] Hege[49] and Fiedler[50] have shown that deposit formation during the heat treatment of milk is strictly correlated with the denaturation of β-LG. The formation of a deposit can be regarded as aggregation. Only activated i.e. unfolded proteins can form a deposit. The greater the concentration of proteins and the higher the value of the heat dependent rate constant the greater will be the rate of deposit formation.

Deposit formation in heat treatment plants can therefore be minimized if the temperature/time conditions during pasteurization are so chosen that no significant denaturation takes place. Thus it is recommended that milk intended for UHT treatment should be pre-heated at 90–95°C for about 120 s. This causes more than 90% of the β-LG to be denatured and no further aggregation will therefore take place on the walls of the heating equipment. Because the walls of the preheater are not hot during the preheating treatment the proteins aggregate with each other and with the casein micelles but do not adhere to any extent on the preheater wall. The question of whether such preheating harms important milk constituents such as thiamin and lysine can be answered with a clear *no* (see Fig. 7 for thiamin, Fig. 8 for lysine). There is no evidence that a preheating treatment of 95°C/120 s of UHT milk affects heat sensitive important milk constituents.

In order to be able to include blood serum albumin and immunoglobulins in the calculations, reaction kinetic data on Cerf[51] are used which are based on the work of Hillier and Lyster.[48] The latter authors investigated the denaturation not only of β-LG and α-LA but also of blood serum albumin. Cerf has based his calculations on the assumption that the denaturation behaviour of blood serum albumin and immunoglobulin fractions is very similar.

The final aim is to present a graph of the denaturation of all of the whey protein fractions. To do this the appropriate calculations are made based on the reaction kinetic data from the denaturation of β-LG A and B, α-LA and BSA and IG. Because the proteose–peptone fraction which represents about 10% of the total whey proteins is regarded as not denaturable the maximum degree of denaturation of the total whey protein fraction can

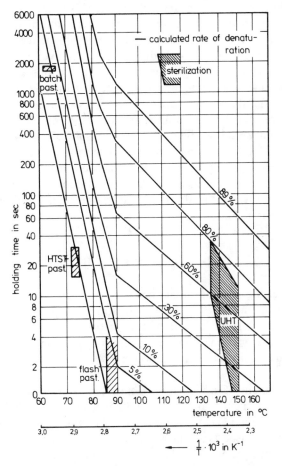

FIG. 19. Calculated degrees of denaturation of whey proteins (β-LG A and B, α-LA, BSA, IG, proteose–peptone) in skim milk.

only be 90%. The concentration of proteose–peptose therefore remains the same whatever the heat treatment. Figure 19 shows the calculated lines of equal degrees of denaturation of the total whey protein fraction.

6. CONCLUSIONS

Table 3 shows, in a simplified form, the changes brought about in milk constituents by heating which are the subject of this contribution. A comparison of the data makes it clear that 'short time' pasteurization is better able to preserve the quality of the milk than 'high' temperature treatment. Milk proteins are already beginning to be denatured by the higher temperatures and this produces a cooked flavour in the milk and leads to an increase in deposit formation in the heating equipment. The shorter holding times during high temperature heating do not compensate for the effects of the temperature differences between the two treatments.

A comparison of UHT treatment and sterilization in the autoclave, on the other hand, presents a different picture. In spite of the higher treatment temperature UHT processing is clearly superior to autoclaving because the treatment times are only 1/1000 of that of autoclaving. Another advantage of UHT treatment is that chemical reactions have lower activation energies than those involved in the destruction of spores. This produces less steep lines as shown in Fig. 2.

7. FURTHER READING

Much interesting work has been done in research establishments throughout the world on the effect of heat treatment on milk. Thanks to the International Dairy Federation and in particular to its group of experts from many countries an extensive bibliography relating to this subject has been established. Further information can be obtained from the following publications.

IDF-Document 157:[58] Burton: Bacteriological, chemical, biochemical and physical changes that occur in milk at temperatures of 100–150°C.

IDF-Document 133 'Monograph on UHT-milk': Teuber and Busse:[59] Microbiological aspects. Renner and Schmidt:[60] Chemical and physico-chemical aspects. Shew:[61] Technical aspects of quality assurance.

IDF-Document 130 'Factors affecting the keeping quality of heat treated milk': Solberg:[62] Enzymatic effects. Busse:[63] Factors of a bacteriological

EFFECTS OF THERMAL PROCESSING ON MILK

Heat treatment	Thermization	Pasteurization		Sterilization	
		Short-time	High	UHT	Autoclave
Temperature (°C)	62–65	71–74	≥85	135–150	≥109–115
Holding time (s)	15–20	>15–30	>2	2–20	$1{\cdot}2$–$24{\cdot}10^3$
Effects	100% ●●●●				
	Very severe ●●●●				
	Severe ●●●				
	Weak ●●				
	Virtually none ●				
				— Not detectable	
				○ Limit of variation	
Killing					
Thermolabile mesophiles	○●●●●	●●●●	●●●	●●●●	●●●●
Thermoduric mesophiles	—	●●●●	●●●	●●●●	●●●●
Mesophillic spores	—	—	—	●●●●	●●●●
Thermophillic spores	●			○	○
Denaturation/ destabilization					
β-Lactoglobulin	—	○	●● ○	○○	●●●●
Whey proteins	—	—	●●	○○	●●●●
Milk proteins inc. casein	—	—	●●●	●●●	●●●
Fat globule membrane	—	—	●●●●	●●●●	●●●●
Vitamin loss					
Thiamin (vit. B_1)	—	—	—	—	●●
Riboflavin (vit. B_2)	—	—	—	—	—
Ascorbic acid (vit. C)	●	●	●	●	●
Binding of amino acids					
Lysine	—	—	—	—	●●
Formation of reaction products	—	—	—	●	●●
Cooked taste	—	—	●	●○	●●●●

nature. Cerf:[64] UHT-treatment. Ashton and Romney:[65] In container sterilisation. Mottar and Naudts:[66] Differences between UHT-milk and in container sterilized milk.

IDF-Bulletin No. 200 'Monograph on pasteurized milk'. Burton:[57] Microbiological aspects. Linden:[67] Biochemical aspects. Deeth:[68] The appearance, texture, flavour and defects of pasteurized milk. Renner[69] and Sieber:[70] Nutritional aspects. Burton:[72] UHT processing of milk.

Special interest to thermal processing of milk is also paid by Lewis[71] and Walstra and Jenness.[35]

REFERENCES

1. NIEMIERSKI, P. Ph. D. thesis. TU Berlin 1986.
2. BRADSHAW, J. G., PEELER, J. T., CORWIN, J. J., HUNT, J. M., TIERNEY, J. T., LARKIN, E. P. and TWEDT, R. M. *Milk. Journal of Food Protection*, 1985, **48**(9), 743–5.
3. VAN DEN BERG, M. G. *IDF B-document No. 97*, 1983.
4. HORAK, F. P. Ph. D. thesis, 1980, Technical University Munich.
5. KESSLER, H. G. *Food Engineering and Dairy Technology*, Publishing House A. Kessler, Freising, FRG, 1981.
6. KIERMEIER, F. *Deutsche Milchwirtschaft*, 1972, **23**, 836–44.
7. PIEN, J. *IDF Monograph on UHT milk*, 1972.
8. PORTER, J. and THOMPSON, S. *IDF Monograph on UHT milk*, 1972.
9. FINK, R. Ph. D. thesis, 1984, Technical University Munich.
10. DANNENBERG, F. Ph. D. thesis, 1986, Technical University Munich.
11. KON, S. K. *Etudes Nutrition No. 17*, Food & Agricultural Organization, Rome (Italy), 1959.
12. NAGASAWA, T., TANAHASHI, T., KUZUYA, Y. and SHIGETA, N. *Jap. J. Zootechn. Sci.* 1960, **31**, 200.
13. HOSTETTLER, H. *Mitt. Geb. Lebensmittelunters. u. Hyg.*, 1965, **56**, 137.
14. GREGORY, M. E. and BURTON, H. *J. Dairy Res.*, 1965, **32**, 13.
15. FORD, J. E., PORTER, J. W. G., THOMPSON, S. Y., TOOTHILL, J. and EDWARDS-WEBB, J. *J. Dairy Res.*, 1969, **36**, 447.
16. KARLIN, R. *Int. Z. Vitaminforsch.*, 1969, **39**, 359.
17. BURTON, H., FORD, J. E., PERKIN, A. G., PORTER, J. W. G., SCOTT, K. J., THOMPSON, S. Y., TOOTHILL, J. and EDWARDS-WEBB, J. D. *J Dairy Res.*, 1970, **37**, 529.
18. LEMBKE, A., FRAHM, H. and WEGENER, K. H. *Kieler Milchwirtschaftliche Forschungsberichte*, 1968, **20**, 331.
19. GÖRNER, F. and UHEROVÁ, R. *Nahrung*, 1980, **24**, (8), 713.
20. BAYOUMI, E. S. Ph. D. thesis, 1981, University of Kiel.
21. THOMAS, E. L., BURTON, H., FORD, J. E. and PERKIN, A. G. *J. Dairy Res.*, 1975, **42**, 285.
22. FINOT, P. A., DEUTSCH, R. and BUJARD, E. *Prog. Food Nutr. Sci.*, 1981, **5**, 345.

23. BLANC, B., FLÜCKIGER, E., RÜEGG, M. and STEIGER, G. *Alimenta Sonderausgabe*, 1980, 27–47.
24. MOTTAR, J. and NAUDTS, M. *Le Lait*, 1979, **59**, 576.
25. TÖTER, D. Ph. D. thesis, 1979, University of Giessen.
26. RENNER, E. and DORGUTH, H. *Deutsche Milchwirtschaft*, 1980, **31**, 505.
27. KONIETZKO, M. Ph. D. thesis, 1981, University of Kiel.
28. ZADOW, J. G. *Aust. J. Dairy Technol.*, 1970, **25**, 123.
29. MORGENSEN, G. and POULSEN, P. R. *Milchwissenschaft*, 1980, **35**, (9) 552.
30. BOSSET, J. O., MARTIN, B. and BLANC, B. *Mitt. Geb. Lebensmittelunters. u. Hyg.*, 1979, **70**, 203.
31. KESSLER, H. G. and FINK, R. *J. Food Sci.*, 1986, **51**, 1105–11.
32. HORAK, F. P. and KESSLER, H. G. *Milchwissenschaft*, 1981, **36** (9) 547.
33. HORAK, F. P. and KESSLER, H. G. *Z. Lebensm. Unters. Forsch.*, 1981, **173**, 1–6.
34. KEENEY, M. and BASSETTE, R. *J. Dairy Sci.*, 1959, **43**, 945.
35. WALSTRA, P. and JENNESS, R. *Dairy Chemistry and Physics*, John Wiley, New York, 1984.
36. LARSON, L. and JENNES, R. *J. Am. Chem. Soc.*, 1952, **74**, 3090–3.
37. GOUGH, P. and JENNES, R. *J. Dairy Sci.*, 1962, **45**, 1033–9.
38. DUPONT, M. *Biochim. Biophys. Acta*, 1965, **102**, 500–13.
39. SAWYER, W. H., NORTON, R. S., NICHOL, L. and MC KENZIE, G. H. *Biochim. Biophys. Acta*, 1971, **243**, 19–30.
40. EL-SHAZLY, A., MAHRAN, G. A. and HOFI, A. A. *Milchwissenschaft*, 1986, **33**, 166–70.
41. DE WIT, J. N. and SWINKELS, G. A. M. *Biochim. Biophys. Acta*, 1980, **624**, 40–50.
42. HARWALKAR, V. R. *J. Dairy Sci.*, 1980, **63**, 1043–51.
43. PARK, K. H. and LUND, D. B. *J. Dairy Sci.*, 1984, **67**, 1699–1706.
44. AGRAWALA, S. P. and REUTER, H. *Milchwissenschaft*, 1979, **34**, 735–7.
45. MOTTAR, J. *Le Lait*, 1981, **61**, 503–16.
46. LARSON, B. L. and ROLLERI, G. D. *J. Dairy. Sci.*, 1955, **38**, 351–60.
47. LYSTER, R. L. J. *J. Dairy Res.*, 1970, **37**, 233–43.
48. HILLIER, R. M. and LYSTER, R. L. J. *J. Dairy Res.*, 1979, **46**, 95–102.
49. HEGE, W. Ph. D. thesis, 1984, Technical University Munich.
50. FIEDLER, J. Ph. D. thesis, 1985, Technical University Munich.
51. CERF, O. *Revue Laitière Francaise*, 1982, **405**, 15–17.
52. DANNENBERG, F. and KESSLER, H. -G. *J. Food Sci.*, 1988, **53**, 258–63.
53. DUPONT, M. *Biochim. Biophys. Acta.*, 1965, **94**, 573–5.
54. KESSLER, H. G. and HORAK, F. P. *Milchwissenschaft*, 1981, **36**, (3), 129–33.
55. WALSTRA, P., VAN DIJK, H. J. M. and GEURTS, T. J. *Neth. Milk Dairy J.*, 1985, **391**, 209–46.
56. SCHOBER, R. and PRINZ, J. *Milchwissenschaft*, 1956, **11**, 466.
57. BURTON, H. *IDF Bulletin No. 200*, 1986.
58. BURTON, H. *IDF Document No. 157*, 1983.
59. TEUBER, M. and BUSSE, M. *IDF Document No. 133*, 1981.
60. RENNER, E. and SCHMIDT, R. *IDF Document No. 133*, 1981.
61. SHEW, D. I. *IDF Document No. 133*, 1981.
62. SOLBERG, P. *IDF Document No. 130*, 1981.
63. BUSSE, M. *IDF Document No. 130*, 1981.
64. CERF, O. *IDF Document No. 130*, 1981.

65. ASHTON, T. R. and ROMNEY, A. J. D. *IDF Document No. 130*, 1981.
66. MOTTAR, J. and NAUDTS, M. *IDF Document No. 130*, 1981.
67. LINDEN, G. *IDF Bulletin No. 200*, 1986.
68. DEETH, H. C. *IDF Bulletin No. 200*, 1986.
69. RENNER, E. *IDF Bulletin No. 200*, 1986.
70. SIEBER, R. *IDF, Bulletin No. 200*, 1986.
71. LEWIS, M. J. In: *Modern Dairy Technology*, (R. K. ROBINSON, Ed.) Elsevier Applied Science Publishers, London and New York, 1986.
72. BURTON, H. *Ultra-High-Temperature Processing of Milk and Milk Products.* Elsevier Applied Science Publishers, London and New York.

Chapter 4

CROSS-FLOW FILTRATION OF BIOLOGICAL FLUIDS ON INORGANIC MEMBRANES: A FIRST STATE OF THE ART

G. M. RIOS, B. TARODO DE LA FUENTE,
M. BENNASAR and C. GUIZARD

*Groupe de Recherche et de Développement des Membranes,
Montpellier, France*

SUMMARY

There is at present increasing use of cross-flow filtration processes in various industrial areas and publication of books or articles on this subject is effusive. Reported works are mainly devoted to traditional processes using high-flux organic asymmetric membranes.

This short monograph does not intend to compete with them by covering the same aspects of cross-flow filtration, but just sets out to present the newly patented inorganic membranes and their related applications in the field of food liquid preservation. Basic mechanisms that control hydrodynamics and mass transport phenomena at wall are investigated with a special attention to the way in which biological fluids and inorganic materials interact, thus modifying separation efficiency. Means are presented that should enable performance to be improved.

In several respects inorganic membranes and associated techniques are yet little more than laboratory curiosities or pure research topics. So pieces of information are few and far between. This article attempts to present a first state-of-art in this up and coming subject.

NOTATION

a	Cross-section area of a single membrane	(L^2)
C	Concentration of solute	(ML^{-3})
d	Diameter	(L)
D	Flow rate	$(L^3 T^{-1})$
$\mathscr{D}$	Solute diffusivity	$(L^2 T^{-1})$
e	Mass boundary layer thickness	(L)
E	Electric field strength	(VL^{-1})
F	Overall volumic concentration factor	—
G	Local volumic concentration factor	—
J	Filtrate flux	(LT^{-1})
k	Mass transfer coefficient	(LT^{-1})
n	Number of membranes in one single stage of unit	—
M	Molar weight	—
P	Pressure	$(ML^{-1} T^{-2})$
r	Poor radius	(L)
R	Hydraulic resistance	$(ML^{-2} T^{-1})$
ε	Filtering surface of a single membrane	(L^2)
S	Total filtering surface in one single stage of unit	(L^2)
t	Time	(T)
T	Solute retention	—
U	Tangential fluid velocity	(LT^{-1})
V	Initial product volume	(L^3)
ΔP	Transmembrane pressure	$(ML^{-1} T^{-2})$
$\Delta \Pi$	Osmotic pressure difference	$(ML^{-1} T^{-2})$
ΔH	Pressure drop or increase	$(ML^{-1} T^{-2})$
θ	Temperature	(K)
v	Surface tension	(MT^{-2})
ψ	Wetting angle	—

Subscripts

a	Pore diameter
b	Bulk solution
c	Channel
e	Pump
f	Filtrate (or permeate)
g	Gel
m	Membrane

p	Product (or freed solution)
r	Retentate (or concentrate)
w	Wall conditions
1,2,X	Stage number

Superscripts

o	Clean membrane properties
+	Dirty membrane complementary properties
*	Instantaneous values
p	Membranes in parallel
s	Membranes in series

1. INTRODUCTION

Filtration is defined as the separation of two or more components from a fluid stream. In conventional usage the term refers to the separation of solid particles. Membrane filtration extends this application further to include the separation of dissolved solutes.

The basic principle of cross-flow filtration is simple as illustrated in Fig. 1. A liquid containing solutes/particles of different sizes flows by a membrane acting as a selective barrier. A hydrostatic pressure is applied to the upstream side which causes solvent plus small components to pass through the membrane by 'capillary diffusion' while large species are retained.

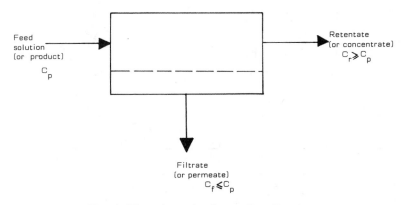

FIG. 1. Floc schematic of cross-flow filtration.

A fluid concentrated in the rejected solutes/particles is collected from the upstream side of the porous wall and a solution of small components is collected from the downstream side.

Retained solute/particle size is a major characteristic distinguishing cross-flow filtration (ultrafiltration, UF, or microfiltration, MF) from other liquid driven pressure separation processes (conventional filtration, CF, or reverse osmosis, RO) as shown in Fig. 2. Other areas of distinction also lie in the nature of basic transport mechanisms, filtering media or fluid flow direction: in reverse osmosis separation is based on differences in solubility and molecular diffusivity of species through the membrane, the so-called 'sorption–diffusion mechanism': in conventional filtration the feed solution flows normally to the filter.

It is common practice to use the term ultrafiltration in those cases where only macromolecules are processed and the term of microfiltration in these other ones where miscellaneous substances or fine particles are involved. In fact, on theoretical grounds the distinction is artificial considering that basic mechanisms are the same in both cases and that no clear correspondence between the size of rejected compounds and the diameter of pores

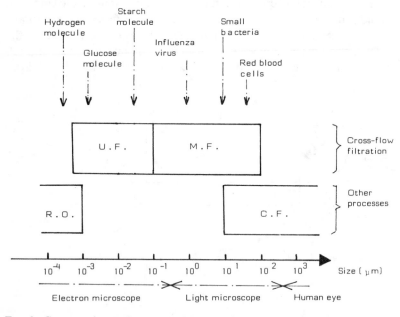

FIG. 2. Comparative performance of liquid driven pressure separation processes.

may be found. In practice membranes with ideal porosity are rarely available and even then changes occur in-service due to solute–membrane and solute–solute interactions. This is especially true with ceramic membranes. For simplicity and conciseness in what follows, the term of cross-flow filtration will be indiscriminately used, and dispersed compounds (molecules, colloids, particles) will be referred to as solutes.

In theory, inorganic membranes should be extremely versatile since they are made of mineral materials and thus should possess a high degree of resistance to chemical and abrasion degradation, should tolerate a wide range of pH and temperature values, thus opening up whole new areas of applications. Most important for biotechnology and food engineering is that they should be repeatedly autoclavable and very stable against microbiological attack. However, they are very expensive compared to traditional organic membranes.

Preparation and structural characteristics of inorganic membranes, modelling of processes, and process design and equipment are successively examined.

2. PREPARATION AND STRUCTURAL CHARACTERISTICS OF INORGANIC MEMBRANES

Many companies and organisations are involved in inorganic membrane research and development. Work on inorganic membranes started with Vycor type glass membranes which were studied in the mid-1940s,[1-3] but exhibited a tendency to dissolve during utilization in filtration processes. Union Carbide (USA) was the first company to produce reliable inorganic membranes in the 1960s; then Euroceral developed alumina membranes from the French National Programme on Nuclear Energy. But the first commercial wholly cross-flow filtration system equipped with inorganic membranes was SFEC (France) in 1978. The CARBOSEP membrane from SFEC was made up of a zirconia microporous layer coated on a macroporous carbon support. New products appeared in 1984 with porous alumina media from Ceraver (France) and Norton (USA). Inorganic membranes are now used to a significant extent in commercial applications meaning that production of crack-free membranes with ultrafine pores and narrow pore size distribution has been achieved. The end of the 1980s will show great emphasis on inorganic membranes and processes.

Inorganic media designed for cross-flow filtration are composed of two parts: the microporous ceramic or metallic support manufactured through powder sintering and the active membrane coated onto the surface of the support, each part exhibiting intrinsic properties. The following deals only with inorganic membranes obtained through firing and sintering of metallic, ceramic or glass particles. 'Dynamic' inorganic membranes formed by filtering colloidal solutions through a porous filter[4, 5] cannot be considered as pure inorganic membranes. The structure of a dynamic membrane is generally made of a hydrous metal oxide trapped in a polymer network. Due to the organic part of the membrane this kind of material does not exhibit the intrinsic properties of inorganic membranes, as mentioned in the Introduction.

2.1. Production of Inorganic Membranes

Although several techniques of generating connected microporosity in inorganic layers have been experienced, powder sintering preserving residual porosity is the most widely used to produce operational membranes. One can observe that micron-sized porous materials for the microfiltration process can be prepared in a classical way through sintering of the appropriate powders. Considering ultrafiltration membranes, only sol–gel techniques can provide ceramic layers with ultrafine pores and a thickness small enough to obtain high flux through the membrane. Coming back to the two parts of an inorganic filtering medium, the supporting part of the membrane has to ensure the mechanical resistance while the supported active layer is designed to provide good separation properties.

2.1.1. Macroporous Support: Characteristics and Elaboration

In addition to a high mechanical resistance the supporting substrate must drain permeate from the membrane without any hydrodynamical resistance. Both mechanical resistance and high flux rate are usually achieved by cylindrical geometry and large grain sintered ceramics. In practice pore diameter for this primary support is about 10 µm. Ordinarily the membranes designed for a specific filtration process have to be coated on the inner part of the tube. Because of the surface rugosity, an ultrafiltration membrane cannot be directly coated on the inner part of the primary support, and a porous layer is necessary between the support and the ultrafiltration membrane. This intermediate layer must exhibit rugosity low enough to allow coating of a thin ultrafiltration membrane, asperities being smaller than the thickness of the membrane to be coated. Pore diameter is about a fraction of micron for this intermediate layer which sometimes can be used directly as a microfiltration membrane. Superposing several layers leads to an asymmetrical structure as represented on Fig. 3. Different fine

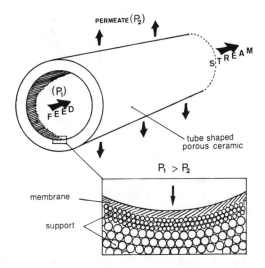

FIG. 3. Schematic texture of an asymmetrical ceramic membrane designed for cross-flow filtration.

powdered materials can be used but corrosion resistance, commercial availability and reasonable sintering temperatures are required for an industrial development. The most common materials used for sintered ceramic supports are: alumina, oxide mixture such as alumina silicate, carbon and carbides, while for sintered metals stainless steel and nickel have been the most widely tried.

The macroporous ceramic supports are elaborated by extrusion of a plastic mixture made of a powder and a binder mixed together in aqueous media. Using different die geometries allows production of tube-shaped or multichannel supports shown in Fig. 4. Multichannel supports offer a higher mechanical resistance and a lower space required/filtering surface ratio than tubular supports. The green substrate obtained through extrusion is dried and fired in order to obtain a sintered material with a residual connected porosity. Concerning the production of an asymmetrical two layered support, it can be achieved either by coextrusion of two powders of different grain sizes in a concentric arrangement, or by coating the primary substrate with the intermediate layer using a slip-casting technique. An example of this category of support is shown in Fig. 5. This alumina support consists of three successive layers exhibiting an asymmet-

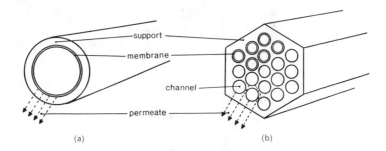

FIG. 4. Schematic cross-sections of tube-shaped (a) and multichannel (b) inorganic
membranes.

rical structure. The external one is 2 mm thick with 10 μm pore diameter,
the internal layers being 15 μm thick with pore diameters varying from
0·7 μm to 0·2 μm.

Elaboration of metallic and carbon supports does not differ significantly.

2.1.2. Inorganic Membranes, Chemical Composition and Coating

Classical powder suspensions can be used for membranes exhibiting pore
diameter at the frontier of the ultrafiltration and the microfiltration
domains; but producing membranes with pore diameter less than 100 nm
requires juxtaposition of colloïdal particles, the pores of the membrane
being generated by residual vacuum between the particles.

The microporous membrane processing concerns the ultrafiltration
domain with pore diameter from 1–2 nm to 100–200 nm. The membranes
are obtained by coating the asymmetrical support with a colloïdal
suspension and firing at a temperature lower than the firing temperature of
the support. Moreover, avoiding important flux decline through the
membrane implies that the membrane thickness does not exceed a few
microns. Related to the foregoing requirements, production of ultrafiltration inorganic membranes is a very convenient application for the
sol–gel process which can allow preparation of coatings and thin layers in
the ceramic field, dense or porous layers being selected depending on the
application.

Two classes of gel have to be considered in sol–gel processing. First, gels
can be formed from colloïdal sols in which gelation is an aggregative effect
between charged particles. These gels exhibit a thixotropic effect used in

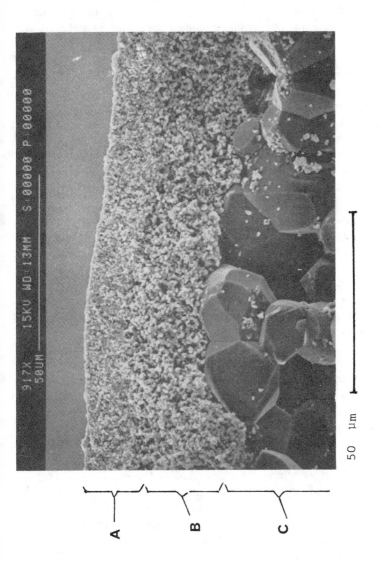

Fig. 5. Section view of an asymmetrical α-alumina membrane from Ceraver (MEMBRALOX). A, internal layer—0·2 μm pore diameter. B, intermediate layer—0·7 μm pore diameter. C, external layer—10 μm pore diameter.

specific coating processes, the layer porosity being directly related to the size and the distribution of the colloïdal particles into the final oxide film. The second class of gels results from an inorganic polymer network formation. Polymerization of metal organic compounds is achieved through hydrolysis and condensation of metal alkoxides. Liquid alkoxides can be easily purified by distillation in order to obtain very pure oxide ceramics, and mixed at the molecular level which enables very precise and varied compositions. Both colloïdal and polymeric gels allow glass and ceramic formation at low temperature.

A general route for membrane preparation using the sol–gel process can be given following the two ways indicated in Fig. 6.

The first example of inorganic ultrafiltration membrane is given by alumina membranes which have been studied by different authors and details on the preparation have been published.[6-8] Concisely, the microporous layer has been obtained working in aqueous media and coating the surface of a macroporous support with an alumina sol. These membranes exhibit narrowly distributed pore diameters from 2·5–10 nm depending on the firing temperature.

More recently zirconia and titania membranes have been elaborated and studied.[9] The authors used zirconium and titanium alkoxides as ceramic membrane precursors and an industrial microporous support from Ceraver shown in Fig. 5. The membranes produced were 3–5 μm thick with the possibility of providing narrow pore size distribution in a 3–10 nm range depending on the firing temperature.

So far in this account, separation processes using the foregoing membranes can be considered as only pressure driven. A new concept in the inorganic membranes area is to design electronic conductive membranes in which the original feature is the possibility of using the active layer both as a filter and as an electrode. With this new class of membranes, selectivity which first depends on pore size distribution, and concentration polarization can be enhanced applying an electrical field in the filtration module. Details on the so-called electro-ultrafiltration process will be given hereinafter. Two ways have been described in the literature to prepare inorganic electro-ultrafiltration membranes.[10, 11] One used a RuO_2–TiO_2 microporous ceramic layer exhibiting metallic conductor behaviour, the other consisted of thin Ni–P and Ni–W–P metallic layers plated on a pre-existing ultrafiltration membrane. The ceramic conductive membrane was produced from $RuCl_3 \cdot 3H_\lambda O$ and $Ti\,(OP_r)_4$ starting materials using a sol–gel process in an organic solvent. Concerning the metal-plated membranes, an electrolysis method was used for plating an alumina membrane with Ni–P and Ni–W–P alloys.

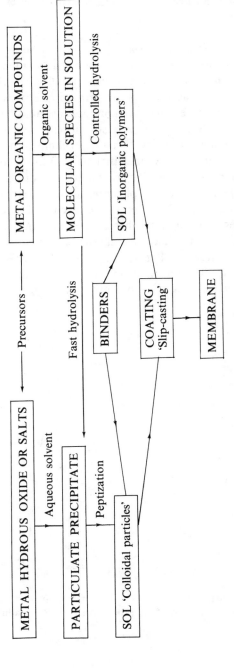

FIG. 6. General route for membrane preparation using the sol–gel process.

2.2. Physico-chemical Properties and Characterization of Inorganic Membranes

2.2.1. *Physico-chemical Properties*

One can see that the filtration flux through a membrane will depend on pore volume and membrane thickness whereas selectivity is related to pore size distribution. On account of preparation methods, the inorganic membranes exhibit a very narrow pore size distribution. Effectively, in membrane manufacturing pore size distribution is directly related to size distribution of ceramic particles. Using well calibrated powers or colloïds generates a very narrow pore size distribution into the formed layer. An important phenomenon during the firing step is the pore size increase under rising sintering temperature. This phenomenon, which is due to the growing of ceramic grains, is generally used to produce membranes with different calibrated pores starting from the same preparation. As an example distributions for a RuO_2–TiO_2 membrane fired at three different firing temperatures are shown in Fig. 7.

Another important feature in the inorganic membranes area is the inter-actions with aqueous media. Organic solvents are completely inert towards

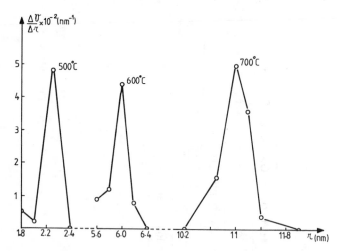

FIG. 7. Pore-size distribution for RuO_2–TiO_2 membranes as a function of firing temperature. v, Membrane porosity. r, Pore radius.

oxides, whereas water molecules can develop strong interactions with oxide surfaces. The hydration mechanism of an oxide surface leads to an amphoteric dissociation of surface MOH groups as follows:

$$M-OH+H_2O \rightleftarrows \begin{array}{l} M-OH_2^+ +OH^- \\ M-O^- +H_3O^+ \end{array}$$

This mechanism explains qualitatively the pH dependence of surface charge and the existence of a pH resulting in a zero net charge, called the isoelectric point (IEP) or zero point charge (ZPC). IEP can be defined as the pH at which the average charge at the surface is zero while ZPC is the pH at which there is no charge at the oxide surface. Several factors can influence IEP values; among these are differences in hydration state, purity and cation radius. With regard to pure water fluxes, Fig. 8 shows an example of two ultrafiltration membranes made of two different oxide materials. Notwithstanding largest pore diameter for the Al_2O_3 membrane, permeation rate is more important for the RuO_2-TiO_2 membrane. If we consider that the pH of pure water was 6·5, close to the ZPC for the Al_2O_3 membrane fired at 1400°C and different from the IEP of the RuO_2-TiO_2 membrane fired at 600°C (close to 5), it can be expected that alumina membrane would retain a superficial layer of more hydrous material which would modify porous volume and pore size in the membrane. These results show it is important to choose oxides for membrane manufacturing as a function of corrosion resistance and mechanical properties but the hydration phenomenon of solid oxides must also be considered.

2.2.2. Characterization of Inorganic Membranes
A number of methods are available to characterize microporous media which can be applied to inorganic membranes and have been discussed elsewhere.[12] The four most usual methods are succinctly presented here to

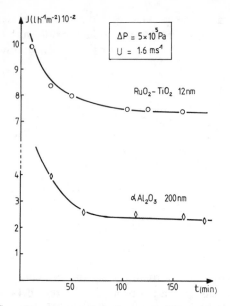

FIG. 8. Water flux drop versus time for two different inorganic membranes (α, Al_2O_3/200 nm and RuO_2–TiO_2/12 nm). Operating conditions: Non-ionized water (pH = 6·2) passed through new membranes well-conditioned by preliminary immersion of several hours at 25°C.

analyse pore size distribution and theoretical permeation flux in a porous inorganic membrane.

Preliminary checking on manufactured inorganic membranes is to disclose eventual cracks in the active layer, which is generally a few microns thick. A very useful method is to control bubble emission through the membrane immersed in a liquid such as an alcohol. The gas pressure at which the bubbles are detected can be related to the size of the cracks using a calibration method.

Taking into account that pore shape in a porous mineral body does not vary under pressure, mercury porometry can be used to measure a large range of pore radii (7·5 μm to less than 2 nm). Pressure of about 4000 bars is required to fill 3-nm micropores. This method using both pore size distribution and porous volume determination is based on mercury penetration inside the pores under pressure. Results can be quantified using

Washburn's equation:

$$P \cdot r = -2v \cdot \cos \psi \tag{1}$$

where P is the mercury pressure above the porous sample, r is the pore radius, v is the surface tension of mercury and ψ is the wetting angle of mercury with the solid. The method offers many advantages: measurement in a few hours of a vast range of pore radii, and available commercial apparatus. But unfortunately it is destructive, the sample being polluted by mercury and no longer unstable for subsequent treatment.

Adsorption and capillary condensation constitute another, very interesting method in which gas adsorption and desorption isotherms on a microporous body serve to give the following parameters:

— specific surface area according to the BET method;[13]
— pore size distribution using the BJH, Montarnal or Pierce method;[13]
— total pore volume;
— general shape of the pores from the shape of the hysteresis curve according to de Boer's classification.

With this method the membranes are not destroyed but the specific range of pore radii measured is limited from 2 nm up to 50 nm.

The last method described here offers a very convenient way of measuring the theoretical permeation flux through a membrane as a function of mean pore radius. The texture and the permeability of the membrane can be investigated by a study of its permeability to a gas, so-called permeametry. For a homogeneous texture the permeability variation with average pressure can be directly related to a texture parameter having a length dimension. If the membrane is likened to a set of straight parallel capillaries, this parameter becomes the mean pore radius r which is the radius of the capillary. Two gas flow modes governing permeability measurement can be investigated:

— stationary pressure conditions during which the gas flows in an established pattern called the steady state;
— transient conditions, in which case decrease in the pressure difference between the two compartments separated by the membrane follows in time.

However, these two flow modes are not easily performed. In the former the pressure regime is slow to stabilize if the gas output is low, while thermal effects can perturb the latter. In both methods the same series of manipulation must be reproduced for each pressure studied and the permeability versus pressure curve is only obtained for several sets of experiments.

3. MODELLING OF PROCESSES

3.1. Polarization and Fouling Phenomena[14, 15]

The accumulation of solutes at the membrane surface is generally the most significant factor which influence the efficiency of cross-flow filtration processes. It results from boundary layer development at the wall.

Whenever a fluid flows past a solid surface, friction between the wall and the liquid causes a retardation of flow in the vicinity of the interface and a speed-up of the fluid layers away from it. The region of lower velocities between the wall and the uniform bulk velocity area is ordinarily referred to as momentum or velocity boundary layer. During cross-flow filtration solutes are brought to the membrane surface by convection transport and a portion of the solvent is removed. This results in a high local concentration zone of partially or completely rejected solutes at the membrane surface. This zone, analogous to the velocity boundary layer but thinner than it, is called the mass or concentration boundary layer for the considered solute (Fig. 9).

This reversible solid build-up also known as 'concentration polarization' induces:

— Changes in physico-chemical properties of the interfacial solution (density, viscosity, solute diffusivity)—for high polarization this may result in gelation at wall.
— Changes in membrane properties due to increased interactions with products (protein adsorption onto alumina membranes, mechanical pore blocking by colloïds). Because of this irreversible 'fouling' phenomenon, solvent flux and solute rejection are irreversibly modified and they will never come back to the original except after complete cleaning.
— Rise in osmotic pressure that partially balances the applied driven pressure.

All these phenomena are responsible for unsteady-state variations of flux and rejection at the start of experiments (Fig. 10), and also cause a steady-state deviation from the simple linear relationship between transmembrane flux and pressure driving force (pressure controlled region) with the flux eventually reaching a constant level where it becomes insensitive to further increase in pressure (mass transfer controlled region) (Fig. 11).

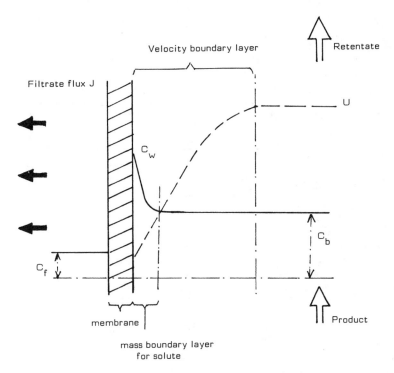

FIG. 9. Schematic of concentration polarization.

To give a good description of process performance one must endeavour to account for all these effects. However, if it is fairly easy to propose sound explanations from them, it appears much more difficult to transcribe information into appropriate analytical and numerical models.

3.2. Model Equations[16–19]

Most of the time the assumption is made that operating conditions are identical at every point of the membrane at any time. On this hypothesis, with pure solvent the flux J through the membrane is given by Darcy's law as a function of the applied pressure, ΔP:

$$J = \Delta P / R^\circ \tag{2}$$

where R° is defined as the clean membrane resistance. In a similar way the flux pressure relationship in the presence of concentration polarization is

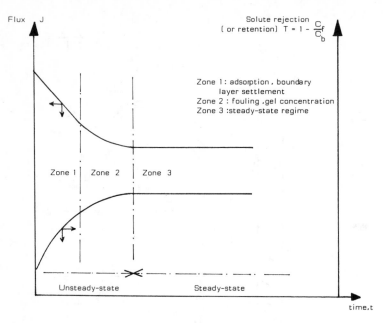

FIG. 10. Flux and rejection versus time.

often described by:

$$J = (\Delta P - \Delta \Pi)/(R^\circ + R^+) \tag{3}$$

where R^+ is the complementary resistance due to fouling or even gelation, and $\Delta \Pi$ the osmotic pressure difference measured for solute concentrations at each membrane solution interface. The physical basis of the observed reduction in ultrafiltrate flux with increased pressure (Figure 11) depends on the extent to which the polarized zone acts, on the one hand as a thermodynamic barrier which reduces the available driven pressure by an increase in $\Delta \Pi$ and on the other hand as a hydrodynamic barrier which offers a high resistance to flow. Considering that for high molecular weight solutes ($M > 100\,000$) concentration must be very high to give effective osmotic pressure and that gels ordinarily form at fairly low concentrations, it may be thought that hydrodynamic limitation more likely occurs. On the contrary when low molecular weight solutes are involved ($M < 100\,000$) thermodynamic limitation may be expected due to higher osmotic pressures and reduced gelation tendency.

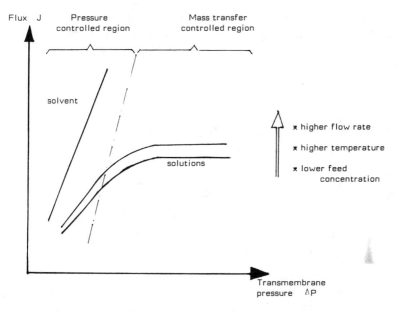

FIG. 11. Generalized correlation between operating parameters and flux.

Another relationship between flux and solute concentrations at wall C_w (C_g in case of gelation), in bulk solution C_b and in the permeate C_f is given by:

$$J = k \cdot \ln\left(\frac{C_w - C_f}{C_b - C_f}\right) \qquad (4)$$

with k the mass transfer coefficient. This equation results from a simple mass balance on the boundary layer. According to the film-theory:

$$k = \frac{\mathscr{D}}{e} \qquad (5)$$

with $\mathscr{D}$ the solute diffusivity and e the corresponding mass boundary layer thickness. For polarized layers in which solute concentrations are sufficiently dilute, k may be considered as a constant. But when limiting flux conditions are reached a sharp decrease in its value may be expected due to viscosity increase of interfacial solution.

Obviously there is a close relationship between C_w/C_f and R^+ values. But at this time there is no equation to predict it. Ordinarily, membrane rejection, or the ability to reject or retain a given solute, is defined as:

$$T = 1 - \frac{C_f}{C_b} \qquad (6)$$

3.3. Factors Affecting Flux and Rejection[20, 21]

3.3.1. Operating Parameters
There are four major operating parameters affecting unit performance: pressure, feed concentration, temperature and flow rate/turbulence.

Pressure: The effect of pressure on flux has already been discussed. Figure 12 shows that in general solute rejection T does not remain constant at elevated ΔP as does J.

Feed concentration: the film-theory model (eqn (4)) states that the flux will decrease exponentially with increasing feed concentration as long as k remains constant. The model has often been used to determine by extrapolation of J versus $\ln C_b$ the gel concentration C_g considering that at $J = 0$, $C_b = C_w$; the slope of J versus $\ln C_b$ is the mass transfer coefficient k. As already explained the problem with the method is that no account for physical property variations at elevated concentration is taken. Variations of T with C_b are much more surprising and yet constitute an entire riddle to be solved (Figs 11 and 13).

Temperature: in general higher temperature will lead to higher flux in both the pressure and mass-transfer controlled regions. This of course assumes there are no other unusual effects occurring simultaneously, such as fouling of the membrane due to precipitation of insoluble substances at elevated ΔP. In the pressure controlled region the effect of θ on flux is mainly due to its effect on fluid viscosity; in the mass transfer controlled region the effect of θ on k through solute diffusivity may also be noticeable. In general considering that inorganic membranes are resistant enough to endure elevated temperatures, it will be best to operate at the highest temperature consistent with the limit of the feed solution (Fig. 11).

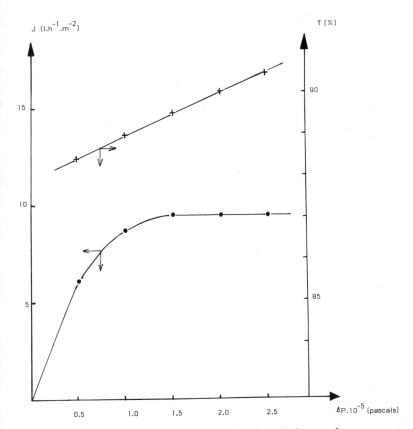

FIG. 12. Cross-flow filtration of gelatin on an alumina tubular membrane: pressure effect (Ceraver; i.d., 15 mm; pore diameter, 0·2 µm; θ, 20°C; pH, 6·7; U, 1 m s^{-1}; C_b, 10 g litre^{-1}).

Flow rate/turbulence: pumping of the fluid has a major effect on flux in the mass transfer controlled region (Fig. 11). Indeed agitation and mixing of the fluid near the membrane surface sweep away the accumulated solutes, thus reducing the thicknesses of boundary layers. This is the simplest and most effective method of controlling the effect of concentration polarization. The magnitude of the effect of flow rate on k and e will depend on whether the flow is turbulent

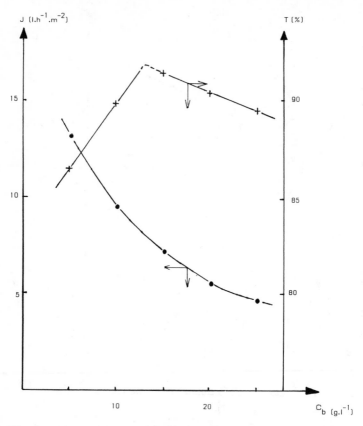

FIG. 13. Cross-flow filtration of gelatin, on an alumina tubular membrane: concentration effect (Ceraver; i.d., 15 mm; pore diameter, 0·2 μm; θ, 20°C; pH, 6·7; U, 1 m s^{-1}; ΔP, 2·10^5 Pa).

or laminar, as well as on rheological properties of the fluid, the key-factor being the shear stress at wall.

Another less common method to effect permeate flux increases is through the introduction of turbulence promoters in the flow conduit. Up to now more attention has been given to fixed promoters due to damage that ordinarily results from movements of free agents at the very fragile surface of traditional organic membranes. In a recent work[22] it has been demonstrated that mineral surfaces are resistant enough to endure the

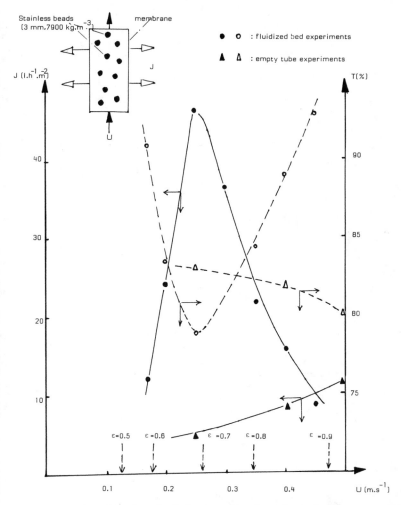

FIG. 14. Cross-flow filtration of gelatin on an alumina tube: turbulence promotor effect (Euroceral; i.d., 27 mm; pore diameter, 0·45 μm; θ, 25°C; pH, 6·0; ΔP, $15 \cdot 10^5$ Pa; C_b, 10 g litre^{-1}).

continuous bombardment of fluidized particles as an example; with this device high permeate fluxes may be obtained at low feed solution velocity with no sharp decrease in solute rejection (Fig. 14).

It is also worth recalling that the flux and the selectivity of cross-flow

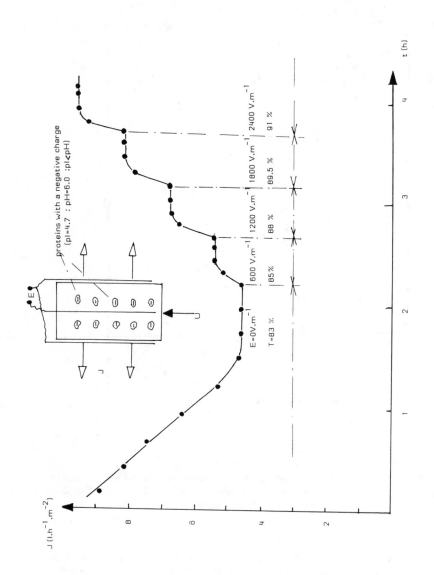

FIG. 15. Cross-flow filtration of gelatin on an alumina tube: electric field effect (Euroceral; i.d., 27 mm; pore diameter, 0·45 μm; θ,

filtration may be improved when processing electrically charged solutes by superposing upon the driven pressure an electric field which acts on the retained solutes to control concentration polarization; as an example[23] with protein solutions at another pH than the one corresponding to isoelectric point, pI, of macromolecules (Fig. 15). From this viewpoint new developments with metallic membranes may be expected in the field of macromolecule separation or even biological reaction.

3.3.2. Design Factors

General design characteristics of membranes are also factors affecting performance (length, shape). More peculiarly the effect of pore diameter must be emphasized.

As already shown, membranes cannot be considered to have a homogeneously permeable surface: pores are relatively sparsely spaced with diameters unevenly distributed. Because of it the radial flow is strongly biased towards the large pores (e.g. 50% of the flow passes through 20–25% of the pores). A corollary to this observation is that loss of the larger pores by plugging or obstruction can cause a major reduction in flux as well as a net change in rejection.

3.3.3. Interactions

The importance of solvent–membrane interactions has already been mentioned (Section 2.2.1). With solutes, similar problems arise.

The effect of product–product interactions may be notable. Two types of systems will be hinted at here, macrosolutes in the presence of microsolutes (ions and pH variation) and macrosolutes in the presence of particulates (colloïds through to suspended solids). For many macrosolutes, properties such as charge, shape, osmotic pressure, diffusion coefficient and viscosity are altered by pH, and ions modify these effects. The presence of particulates that can either retard solute back-diffusion at wall (such as fine colloïds) or enhance turbulence thus reducing polarization (as with suspended solids) is also worth considering. These phenomena ordinarily lead to decrease or increase in flux and rejection.

Product–membrane interactions have also a major effect with regard to performance. Because of it proteins with a molar weight as low as 100 000 daltons (less than 10 nm) may be satisfactorily rejected (80%) using an alumina membrane with a nominal mean pore diameter as high as 0.2 μm! There the role of the protein deposited layer at the membrane surface, the real 'narrow dynamic membrane' blocking macromolecules, is considerable.

4. PROCESS DESIGN AND EQUIPMENT

Distinction can be made between two types of applications according to the substance processed, the separation sought and the fraction recovered:

— Selective or total concentration treatments for substances such as milk and milk derivatives (whey, coagulum, buttermilk, etc.), blood (plasma), soybean milk, eggwhite, microorganisms, effluents, etc. in which the concentrate is the sought fraction.
— Clarification and/or chemical and/or microbiological stabilization treatments for fruit and vegetable musts, as well as juices, alcoholic beverages (wine, beer, cider), vinegar, oil, brine, products of biological reactions, effluents, etc. in which the filtrate is recovered.

As examples concentration of milk coagulum in cheese-making, and the clarification and stabilization of wine in oenology are described in detail. Then, other potential applications of mineral membranes in the food industry are briefly discussed.

4.1. Dairy Industry—Lactic Curd: an Example of Continuous Operation for Concentration

4.1.1. Cross-Flow Filtration of Dairy Products
Figure 16 points out the main uses of cross-flow filtration in milk processing and cheese or butter-making with:

MF-1: cold skimming and pasteurization of milk
MF-2: recovery of fat from buttermilk
UF-1: partial pre-concentration of cheese milk and standardization of the proteins of market milk using the filtrate recovered
UF-2: partial pre-concentration of cheese milk seeded and/or renneted
UF-3: draining of curd (cheese or cream cheese)
UF-4: recovery of proteins from whey
UF-5: prior draining of milk producing liquid pre-cheese.

The draining of lactic curds in the manufacture of fresh cream cheese is examined.

4.1.2. Draining of Curd [24, 25]

4.1.2.1. *Definition.* In curds obtained by chemical acidification (lactic acid) of skim or whole skimmed milk, it is the separation of an aqueous

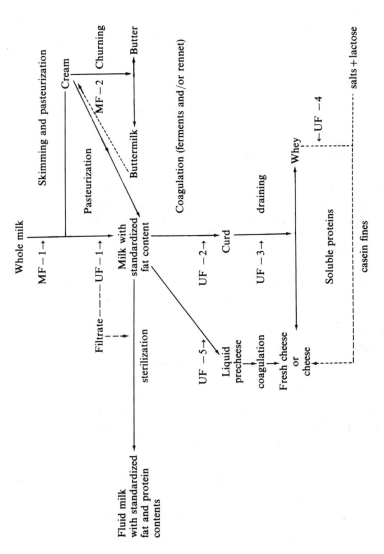

FIG. 16. Use of mineral membrane cross-flow filtration techniques in milk, cheese and butter production.

phase containing only lactose and minerals (filtrate) from a semi-solid phase containing all the proteins (particularly the soluble ones) and fat (retentate). At the process discharge, the total dry matter content of the retentate is the one desired for the finished product (cream cheese).

Up until today, this operation was mainly carried out using centrifugal separators. Figure 17 shows the diagram of a typical cross-flow filtration installation.

4.1.2.2. Choice of membrane. The membrane must have as high a filtrate flow rate as possible with maximum retention of protein and fatty substances.

In previous works,[26, 27] we demonstrated that limitation in filtration performance more often results from interactions that develop between the fluid constituents and the membrane material. With milk, these interactions lead to the formation of a porous deposit, both resulting from piling up of casein molecules and progressive fouling by soluble proteins (α-lactalbumin, β-lactoglobulin, and bovine serum albumin) of the cake so-formed; calcium and phosphorus salts also contribute to the process (Fig. 18).

Similar studies carried out with lactic curd on Membralox alumina membranes with pore diameters (d_a) of 0·2 and 0·8 μm showed that fouling more strongly affects the 0·8 μm filter performance than the 0·2 μm one. The work also showed the advantages of using membranes with a relatively high hydraulic diameter to reduce head losses when the curd concentrates and reaches high dry matter contents. We therefore choose multichannel membranes exhibiting 0·2 μm pore diameter and 2 or 3 mm channel radius (see Fig. 4) with a length equal to 800 mm.

4.1.2.3. Selection of operating conditions. The effect of operating parameters (trans-membrane pressure, tangential velocity, temperature of the substance) on filtrate flux and protein retention as a function of time and concentration of retentate, was tested on a single membrane. The following optimal conditions were determined:

Pressure: Flux linearly varies with pressure up to approximately 5 bars, then it becomes independent of it (cf. Fig. 11) — retention is also close to maximum from 5 bars onwards. Operating at a higher pressure is therefore not worthwhile, it would lead to increased power consumption.

Velocity: Flux improves with velocity (cf. Fig. 11) with no notable loss in rejection. The limiting factor remains increased head loss. With

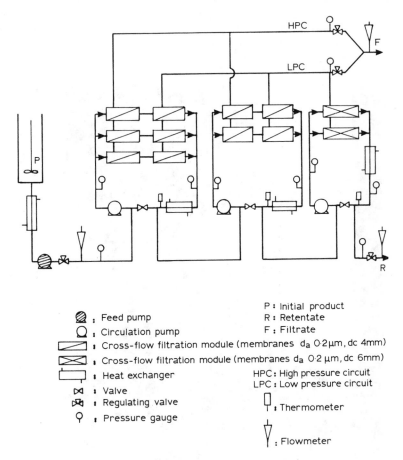

FIG. 17. Diagram of an industrial installation for draining curds.

a velocity of 4 m s^{-1}, pressure drop is limited to 3 bars ($d_c = 4$ mm) or 2 bars ($d_c = 6$ mm) with a final dry matter content of 200 g kg^{-1}.

Temperature: Filtrate flow rate increases with temperature (cf. Fig. 11) but protein retention tends to fall. A temperature of 50°C, which is compatible with the organoleptic qualities of the product, seems to be a good choice, with retention of over 95% of protein material.

Filtrate flux and protein retention decrease when the concentration of

FIG. 18. Scanning electron microscopy observation of deposit formed after contacting reconstituted skim milk (1·0% mV) with a MEMBRALOX membrane (0·2 µm): m; casein micelles; Al, alumina particles.

retentate increases. Steady state is reached more quickly and the effect of operating parameters is less sensitive.

The method used to prepare curd affects filtration performance. Curd produced using biological methods gives better flux but poorer protein retention than curds acidified chemically. Curd from renneted or fat-enriched milk generally gives better performances.

Cleaning and disinfection of membranes is carried out with soda (2–3%, 70–80°C, 30 min), nitric acid (1–2%, 60–70°C, 20 min) and bleach (1% of a 48 chlorometric degree solution of sodium hypochlorite, 20–30°C, 10 min) alternated with water rinsing steps (preferably hot water). For curd, it is advisable to schedule a cleaning–disinfection cycle every 10–12 hours. The type of cleaning, the choice of disinfection agents and their use ensure simplicity, effectiveness and economy.

Figure 19 shows performance for a $4 \times$ concentration operation of curds prepared using biological or chemical methods from renneted skim milk; this is the case when manufacturing fresh cream cheese with 0% fat content and 21% total dry matter content. It appears that an average filtrate flux of 120 litres h^{-1} m^{-2} is obtained, and protein retention is as high as 95–96%. Pressure drop equals 1 bar with the initial product and 3 bars with the final product using 4 mm channel radius. Under the same working conditions it remains less than 2 bars with 6 mm radius.

4.1.2.4. Sizing and operation of an industrial plant. The following points need to be carefully taken into account in the design and working of an industrial plant:

(1) Optimal operating conditions with regard to flux and retention objectives as previously mentioned.
(2) Amount of material, V, to be processed in a given time t, and final concentration of retentate.
(3) Membrane characteristics (filtration surface and cross-section area for fluid flow).

These all at once govern the total filtration surface and the stage number of unit.

As indicated in Fig. 17, each stage is a loop with a recirculating pump setting up tangential velocity and a heat exchanger maintaining product temperature. The product is concentrated progressively in each loop and reaches its final concentration at the discharge of the last stage.

Let F be the concentration factor for the whole plant expressed as a function of overall product D_p, retentate D_r and filtrate D_f flowrates as in the

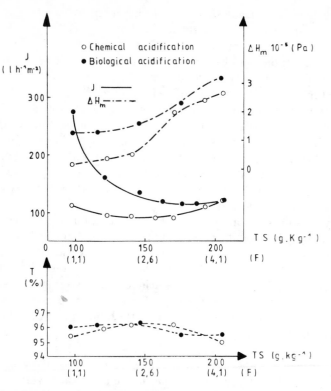

FIG. 19. Ultrafiltrate flux, head loss and protein retention rate as a function of the total dry matter content of a biological or chemical coagulum (pH = 4·40) treated on a MEMBRALOX membrane (d_a, 0·2 μm; d_c, 4 mm) at $\Delta P = 5 \cdot 10^5$ Pa, $U = 3 \cdot 8$ m s^{-1} and $\theta = 50°$C. TS, total solid; F, concentration factor.

equation:

$$F = \frac{D_p}{D_r} \text{ where } D_p = D_r + D_f = \frac{V}{t} \tag{7}$$

The total filtrate flowrate of the installation may be written:

$$D_f = D_p \cdot \frac{F-1}{F} \tag{8}$$

Let F_1, F_2, F_x be the over volumic concentration factors at discharges of stages 1, 2, X; similarly defined as:

$$F_X = \frac{D_p}{D_{rX}} = \frac{D_p}{D_p - (D_{f1} + \cdots + D_{fX})}$$ (9)

with D_{f1}, D_{f2}, D_{fX} the filtrate flow rates for stages 1, 2, X.

Let J_1, J_2, J_X be the filtrate fluxes in stages 1, 2, X and J the mean flux for the whole installation.

Let S_1, S_2, S_X be the membrane surface of stages 1, 2, X and S the total surface area of the plant:

$$D_f = JS = J_1 S_1 + J_2 S_2 + \cdots + J_X S_X$$ (10)

Knowing J_1 from the preliminary study of operating conditions S_1 will be deduced as follows from a fixed F_1:

$$S_1 = \frac{D_p}{J_1} \cdot \frac{F_1 - 1}{F_1}$$ (11)

Likewise, the area of stage X will be related to J_X according to:

$$S_X = \frac{D_p \cdot (F_X - 1) - F_X \cdot (D_{f1} + \cdots + D_{fX-1})}{J_X \cdot F_X}$$ (12)

If, instead of considering the overall volumical concentration factors at the discharges of the various stages, the concentration factors for each stage G_1, G_2, G_X are considered in the form of:

$$G_X = \frac{D_{pX-1}}{D_{pX-1} - D_{fX}}$$ (13)

the membrane surface for X-stage becomes expressed as:

$$S_X = \frac{D_p}{J_X \cdot G_1 \cdot G_2 \ldots G_{X-1}} \cdot \frac{G_{X-1}}{G_X}$$ (14)

Since the surface of a single membrane is s, the number of membranes required in stage X is:

$$n_X = \frac{S_X}{s}$$ (15)

The arrangement of membranes in each stage will depend on both the tangential velocity U and the pressure drop in every membrane ΔH_m, in relation to the characteristics of the recirculating pumps (flux, heat) and the total cross section area for a single membrane a.

With a pump ensuring a maximal flow-rate D_e and a total head equivalent to ΔH_e, it will be possible to install in parallel a maximum number of membranes, n^p, equal to:

$$n^p = \frac{D_e}{a \cdot U} \tag{16}$$

and in series:

$$n^s = \frac{\Delta H_e}{\Delta H_m} \tag{17}$$

Passage of the largest quantity of filtrate can be carried out most easily in the first stage.

The initial product with the smallest solid load is the most sensitive to the effect of the velocity; which results in the least pressure drop. There will thus be more membranes installed in series and in parallel in this first stage, and a centrifugal recirculating pump with a high flux and low head loss will be chosen.

In the last stage, also called the finisher, the already very viscous product will be brought to final concentration F by removing a small amount of filtrate. Ordinarily this stage consists of a small number of membranes mounted in parallel rather than in series and fitted with a 6 mm membrane d_c to reduce head loss. The recirculating pump has a smaller flux; it can be centrifugal or volumetric if considerable head losses are to be overcome.

The operating pressure of the whole 'installation is provided by a variable flux centrifugal feed pump, characterized by a low flux ($D_e > V/t$) and a high ΔH_e.

The concentration process may be controlled by means of regulating valves on product, retentate and filtrate pipes. Control of operating pressure in every stage is performed using counter pressure valves on the retentate and filtrate circuits (high and low pressure). Concentration in the retentate C_r depends on the initial concentration C_p and the overall volumic concentration factor F. It may be expressed as:[28]

$$C_r = C_p + C_p \cdot T \cdot (F - 1) \tag{18}$$

on the hypothesis that the rate of retention (T) does not vary from one stage to the other.

4.1.2.5. Advantages of the technique. Traditionally, the draining of curd is carried out industrially using a centrifugal separator. Tangential filtration makes it possible to recover a large proportion of the soluble proteins lost in the centrifugation serum and to increase cheese yield. Sometimes, improved texture was observed at the equivalent total dry matter content. Mineral membranes also have the advantage of being able to process curd directly, improving the performance of the operation and avoiding bitterness in finished products in comparison with filtration of milk on polymer membranes before acidification.[29]

4.2. Wine Industry – An Example of Quasi-continuous Operation for Clarification and Stabilization

4.2.1. Cross-Flow Filtration in the Wine Preparation Process
The points at which filtration is used in oenology are shown in Fig. 20.

MF-1:   clarification and biological stabilization of musts
MF-2:   clarification and biological stabilization of unprocessed wines

The latter case is examined here.

4.2.2. Clarification and Stabilization of Wines[30,31]

4.2.2.1. Definition of the process. It is the separation from unprocessed wines (red, 'rosé', white) of a clear, brilliant phase which is chemically and biologically stable (filtrate) and a semi-solid phase made up of particles, colloidal material in suspension and bacteria (retentate). The process is continued until a pelletable residue is obtained which represents only 1–2% of the initial product.

Traditionally this operation is carried out in industry by fining, successive filtration operations on finer and finer filters (diatomite, plates), chemical stabilization by refrigeration and biological stabilization, either by pasteurization using a heat exchanger or by frontal filtration on polymer membranes. Figure 21 shows the schematic diagram of an industrial cross-flow filtration unit for clarifying and stabilizing wine. This process is a continuous one as regard to fluid circulation; but retained solids are recirculated. That is the reason why it is said to be 'quasi-continuous'.

4.2.2.2. Choice of membrane. The membrane must allow as high a filtrate flux as possible together with minimum retention of aromatic, colouring

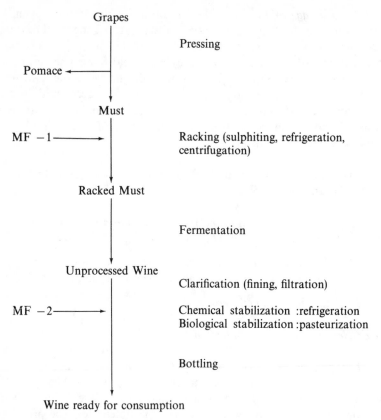

FIG. 20. Role of mineral membrane cross-flow filtration techniques in oenology.

and gustative substances; on the contrary retention of particles and colloids responsible for turbidity, as well as of microorganisms (particularly yeasts) must be excellent.

Our work on Euroceral and MEMBRALOX alumina membranes with various pores ($d_a = 0.2$, 0.45, 0.80, 1.2 and 1.5 μm) and diameters ($d_c = 4$ and 15 mm), led us to select a MEMBRALOX membrane, $d_a = 0.2$ μm, $d_c = 4$ mm. This membrane gave one of the highest fluxes and ensured a perfect bacteria retention as well as a maximum filtrate brilliance. No notable modification of the organoleptic features resulted, as indicated by comparative sensorial analysis.

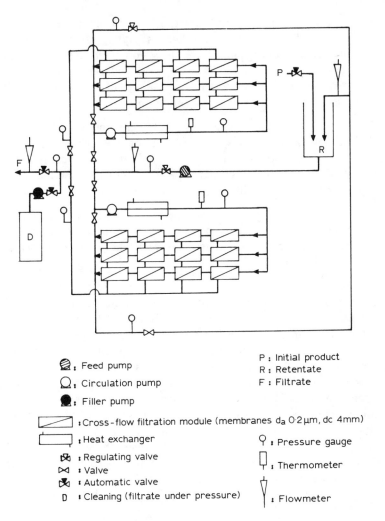

    🌀 **:** Feed pump

    ◯ **:** Circulation pump

    ⬤ **:** Filler pump

    ▱ **:** Cross-flow filtration module (membranes d_a 0·2 μm, d_c 4mm)

    ▭ **:** Heat exchanger

    ⬤ **:** Regulating valve

    ◄► **:** Valve

    ⬤ **:** Automatic valve

    D **:** Cleaning (filtrate under pressure)

P **:** Initial product
R **:** Retentate
F **:** Filtrate

♀ **:** Pressure gauge

▯ **:** Thermometer

▽ **:** Flowmeter

FIG. 21. Diagram of an industrial installation for clarifying and stabilizing wine.

4.2.2.3. Choice of operating conditions. As above, the effect of operating conditions on flux, clarification (measured by a classical filterability index)[30] and the total bacteria counting in the filtrate were studied on a test-membrane as a function of time and concentration of retentate.

The following optimal conditions were determined:

Pressure: Flux levels off from 3 bars (cf. Fig. 11). Bacteria are absent from the filtrate at all pressures, but at high pressures (>7 bars), modification of organoleptic features is observed after overclarification. A pressure of 3 bars is therefore sufficient to ensure good performance with respect to the qualities of the finished product without increased energy consumption.

Velocity: Performances increase with velocity (cf. Fig. 11) without notable modification of the chemical or biological quality of the filtrate. Since pressure drop remains low for this type of process and product it can be envisaged to operate at a circulation velocity of the order of $5 \, \mathrm{m \, s^{-1}}$ with $\Delta H_m < 1$ bar.

Temperature: Although it improves performance (cf. Fig. 11), heating of wine is not permitted because it is harmful to organoleptic characteristics. Sensorial tests showed that product temperature should not exceed 20°C.

The effect of concentration is less distinct than in the previous case: clarification and sterility of the filtrate are not affected; there is only a flux tendency to decrease when concentration increases. Clarification pretreatment of wine (fining, diatomite filtration) significantly improve flux without reducing retention characteristics; however, some of the advantages of cross-flow filtration are removed.

The wine-making method is important since wines made using the 'maceration carbonique' technique (fermentation under regulated pressure using bubble caps) do not filter as well as those made using traditional methods with finished products of comparable quality. The type of wine also affects performances. Fluxes are better when the wine is less highly coloured (white > rosé > red). Sterility of the finished product is the same in all cases. Cleaning and disinfection are carried out as for dairy products with soda, nitric acid and bleach, but one cycle every 24 h is generally sufficient.

Figure 22 shows operating flux and retention during the clarifying and stabilizing of unprocessed red wine for 50 hours. Retention of total dry matter during filtration did not exceed 10%, clarification was efficient and the filtrate was sterile. The mean flux of filtrate (wine) was of the order of 50 litres $\mathrm{h^{-1} \, m^{-2}}$ for the whole duration of operation. The final residue represented 1·4% of the total processed fluid; it contained 180 g litre $^{-1}$ dry matter and was liquid in appearance. There was little variation of pressure drop (0·8–1 bar) during operation.

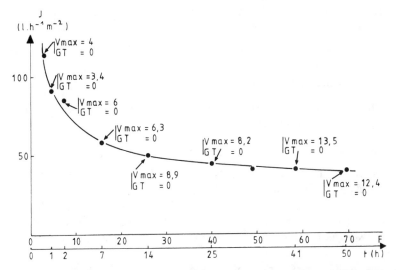

FIG. 22. Ultrafiltrate flux, filtrability index and total bacteria as a function of filtration time for a red wine (unprocessed characteristics: filtrability index $V_{max} = 0.4$ litre; total bacteria $= 10^5/100$ ml) treated on a MEMBRALOX membrane (d_a, 0·2 μm; d_c, 4 mm) at $\Delta P = 5 \cdot 10^5$ Pa, $U = 4 \cdot 7$ m s^{-1} and $\theta = 20°$C. GT, germ total 1. The filtrability index V_{max} must be at least 41 to ensure satisfactory filtration of wines.

4.2.2.4. Sizing and operation of an industrial plant. The same features as the ones already discussed for curd should be taken into consideration in the design of an industrial installation. In a clarification operation, the product filtered is discharged continuously whereas the retentate is fed into the supply tank again. The concentration of retentate in the filter circuit must remain as low as possible in order to maintain the performance level. Filter stages are thus installed in parallel and the retentate/filtrate flowrate ratio is imposed in each circuit as well as in the whole installation so as to keep to a minimum the instantaneous concentration factor F^*. F^* is the same in each circuit and in the installation. In general, $D_r^*/D_f^* > 4$, corresponding to

$$F^* = (D_f^* + D_r^*)D_r^* < 1 \cdot 25.$$

Let the quantity of wine to be clarified in time t be V, with an average filtrate flux J, corresponding to a concentration factor $F^* = 1 \cdot 25$ in the circuit. Let F be the overall volumic concentration factor to be attained at the end of the process. The hourly flow rate of clarified wine is:

$$D_f = \frac{V}{t} \cdot \frac{F-1}{F} \simeq \frac{V}{t} \qquad (19)$$

with a sufficiently large $F(F > 90)$.

The total membrane surface required is $S = D_f/J$, i.e. the following number of membranes with a unit surface area of s:

$$n = S/s$$

These n elements are distributed in several identical stages installed in parallel and each consisting of a recirculation pump and a heat exchanger. The number of stages and the layout of the membranes in each one is a function of the tangential velocity and the head losses. Because head losses are lower than with curd, a greater number of membranes in series can be planned in each filter loop.

Operating pressure throughout the installation is always imposed by a high pressure centrifugal feed pump: its variable flux (D_e) must nevertheless be sufficiently great in this case to ensure that retentate is recycled under the ideal conditions mentioned above: $F^* < 1\cdot25$, i.e. $D_e > D_p^*$, in which

$$D_p^* = \frac{F^*}{F^* - 1} \cdot D_f^* \simeq 5D_f^* \simeq 5\frac{V}{t} \qquad (20)$$

finally giving: $D_e > 5\dfrac{V}{t}$

Concentration in the filtration circuit is controlled by adjustment of feed flow and retentate flow rate with regulation valves, it being known that in this case the volumic concentration factor $F^* = D_p^*/D_r^*$ must not exceed $1\cdot25$.

Operating conditions, and in particular the working pressure, are controlled in each stage by counter-pressure regulation valves on the retentate circuits. The important point in this case is to ensure that there is sufficient pressure at every stage exit. Indeed, the performances are little affected by the rise in pressure caused by losses of head and separate regulation systems are not required on the filtrate circuits. An automated system of continuous intake of fresh wine slows the increase in concentration in the feed tank and slows the fall in performance. A counter-current flushing system returns a small amount of filtrate through the membrane when transmembrane pressure becomes excessive.

This system is made possible by the strength of the mineral membranes and by the good bonding of filter layer to the support: it limits the reduction in the filtrate flux in times and makes it possible to leave longer intervals between cleaning cycles.[32]

4.2.2.5. Advantages of the technique.   Unlike the traditional clarification with biological and chemical stabilization processes mentioned above, cross-flow filtration produces clarified, sterile wine ready for bottling, in a single operation. Its advantages are thus simplicity, continuity and rapidity. In addition, cross-flow filtration enables precipitation of tartar in wine to be carried out faster and at a higher temperature, thus saving energy.[33]

Mineral membranes require less consumables then the polymer membranes used for the frontal sterilization filtration of wine.

4.3. Other Various Fields of Application of Cross-Flow Filtration

Recent work on various applications of mineral membranes in the food industry is covered briefly here.

Supply of potable water for food industries can be by ultrafiltration of non-potable fresh water on membranes with a pore diameter of 6 nm.[34] In oenology, in additon to the work mentioned above, other researches have been carried out on the clarification and stabilizing of wines and also of musts.[35,36] Cross-flow microfiltration is also used to clarify and stabilize beer and for the recovery of extracts from brewery tank bottoms.[37] The process has also similar advantages in fruit juice processing, as an example to valorize kiwifruit sorting rejects by manufacture of juice.[38]

The dairy industry is a major field for the use of membrane processes. Various types of cheese can be manufactured using processes incorporating cross-flow filtration on mineral membranes (soft cheese,[39] pressed cheese,[40] goat cheese.[41] The technique can also be used in cheeses of specific origin for cold pre-concentration of renneted raw milk.[28,42] Membranes with pore diameter 1·8 μm are used for skimming and cold pasteurizing of whole raw milk.[43] Cheesemaking milk whey can be processed by clarification, removal of bacteria[44] and soluble proteins recovery.[45]

The technique can also be used to recover plasma proteins from bovine blood.[46] Finally, mineral membranes can be used throughout the whole biotechnology sector, in particular because they can be steam-sterilized.[47]

5. CONCLUSIONS

Following a brief survey of inorganic membrane manufacturing and properties, and a short presentation of basic transport phenomena that control filtration, a design strategy of cross-flow filtration units either using tubular or multichannel ceramic membranes has been proposed. Two applications have been more particularly examined: concentration of milk coagulum in cheese-making and clarification/stabilization of wine in oenology. The main advantages of the technique with regard to other more traditional processes have been clearly emphasized: simplicity, continuity and rapidity. Energy is minimized with unaltered and even sometimes improved product quality. Other actual and potential applications have been also mentioned. Research in the near future should tend to patent yet more efficient membranes and to elaborate original related processes in the most widely spread areas.

REFERENCES

1. NORDBERG, M. E. Properties of some Vycor-Brand glasses. *J. of the American Ceramic Society*, 1944, **27**, 299.
2. BALLOU, E. V. and WYDEVEN, T. Solute rejection by porous glass membranes. II. Pore size distributions and membranes permeabilities. *J. Colloid Interface Sci.*, 1972, **41**, 198.
3. SCHNABEL, R. and VAULOUT, W. High pressure techniques with porous glass 'membranes. *Desalination*, 1978, **24**, 249.
4. KRAUS, K. A., SHOR, A. J. and JOHNSON, J. S. Jr. Hyperfiltration studies. X. Hyperfiltration with dynamically formed membranes. *Desalination*, 1967, **2**, 243.
5. JOHNSON, J. S., Jr. MINTURN, R. E. and WADIA, P. H. Hyperfiltration. XXI. Dynamically formed hydrous Zn (IV) oxide–polyacrylate membranes. *J. Electroanal. Chem.*, 1972, **37**, 267.
6. LEENAARS, A. F. M., KEIZER, K. and BURGGRAAF, A. J. The preparation and characterization of alumina membranes with ultrafine pores. Part 1. Microstructural investigations on non supported membranes. *J. Mater. Sci.*, 1984, **19**, 1077.
7. LEENAARS, A. F. M. and BURGGRAAF, A. J. The preparation and characterization of alumina membranes with ultrafine pores. Part 2. The formation of supported membranes. *J. Colloid Interface Sci.*, 1985, **105**, 27.
8. LARBOT, A., ALARY, J. A., GUIZARD, C., COT, L. and GILLOT, J. New inorganic ultrafiltration membranes: preparation and characterization. *Int. J. High Technology Ceramics*, 1987, **3**, 145.
9. LARBOT, A., FABRE, J. P., GUIZARD, C., COT, L. and GILLOT, J. New inorganic

ultrafiltration membranes. Part 1. Titania membranes. Part 2. Zirconia membranes. *J. American Ceramic Society*, in press.

10. GUIZARD, C., IDRISSI, N., LARBOT, A. and COT, L. An electronic conductive membrane from a sol–gel process. *British Ceramic Proceedings*, 1986, **38**, 263.

11. GUIZARD, C., LEGAULT, F., IDRISSI, N., LARBOT, A., COT, L. and GAVACH, C. Electronic conductive mineral membrane designed for electro-ultrafiltration. *J. Memb. Science*, in press.

12. RASNEUR, B. Porosimetry: characterization of porous membranes. *Proceedings of the Summer School on Membrane Science and Technology*, September 3–7, 1984, Cadarache, France.

13. JOUERRE, C. E.. Traité de céramiques et matériaux minéraux, edn. septima, Paris, 1980.

14. CHEYRAN, M. *Ultrafiltration Handbook*, Technomic Publish. Comp. Inc., Lancaster, USA, 1986.

15. BELFORT, G. *Synthetic membrane processes*. Academic Press Inc., London, New York, Tokyo, 1984.

16. VILKER, V. L., COLTON, C. K., SMITH, K. A. and GREEN, D. L. The osmotic pressure of concentrated protein and lipoprotein solutions and its significance to ultrafiltration. *J. Memb. Science*, 1984, **20**, 63.

17. WIJMANS, J. G., NAKAO, S. and SMOLDERS, C. A. Flux limitation in ultrafiltration: osmotic pressure model and gel layer model. *J. Memb. Science*, 1984, **20**, 115.

18. NAKAO, S., NOMURA, T. and KIMURA, S. Characteristics of macromolecular gel layer formed on ultrafiltration tubular membrane. *AIChE Journal* 1979, **25**, 615.

19. AIMAR, P. and SANCHEZ, V. A novel approach to transfer limiting phenomena during ultrafiltration of macromolecules. *Ind. Eng. Chem. Fundam.* 1986, **25**, 789.

20. INGHAM, K. C., BASBY, T. F., SAHLESTROM, Y. and CASTINO, F. Separation of macromolecules by ultrafiltration: influence of protein adsorption, protein–protein interactions and concentration polarization. In *Ultrafiltration: membranes and applications* (ed. by A. R. Cooper), Plenum Press, New York and London, *Polymer Science and Technology*, Vol. 13, 1980, p. 141.

21. FANE, A. G. Factors affecting flux and rejection in ultrafiltration. *J. Sep. Proc. Technol.*, 1983, **4**, 15.

22. RIOS, G. M., RAKOTOARISOA, H. and TARODO DE LA FUENTE, B. Basic transport mechanisms of ultrafiltration in the presence of fluidized particles. *J. Memb. Science*, 1987, **34**, 331.

23. RIOS, G. M., RAKOTOARISOA, H. and TARODO DE LA FUENTE, B. Basic transport mechanisms of ultrafiltration in the presence of an electric field. *J. Memb. Science*, in press.

24. ATTIA, H. Ultrafiltration du lait, de laits acidifiés et des caillés lactiques sur membrane minérale, Ph. D. thesis, 1987, Montpellier, France.

25. ATTIA, H., BENNASAR, M., and TARODO DE LA FUENTE, B.Ultrafiltration sur mémbrane minérale de laits acidifiés à divers pH par voie biologique ou chimique et de coagulum lactique. *Le Lait*, 1988, **68**(1), 13.

26. VETIER, C., BENNASAR, M., and TARODO DE LA FUENTE, B. Etude des

interactions entre constituants du lait et membranes minérales de microfiltration. *Le Lait*, 1986, **66**, 269.

27. VETIER, C., BENNASAR, M., and TARODO DE LA FUENTE, B. Study of the fouling of a mineral microfiltration membrane using SEM and physico-chemical analysis in the treatment of milk. *J. Dairy Research*, 1988, **55**(3), 381.

28. BENNASAR, M. and TARODO DE LA FUENTE, B. Ultrafiltration modérée de lait cru entier froid sur membranes minérales: essais semi-industriels. *Le Lait*, 1983, **63**, 246.

29. MAHAUT, M., MAUBOIS, J. L. M., ZINK, A., PANNETIER, R. and VEYRE, R. Eléments de fabrication de fromage frais par ultrafiltration sur membranes de coagulum de lait. *Tech. Lait*, 1982, **961**, 9.

30. POIRIER, D. Contribution à l'étude de la microfiltration tangentielle sur membranes minérales et de son application á la clarification et la stabilisation des vins, Ph. D. thesis. 1985. Montpellier, France.

31. POIRIER, D., MARIS, F., BENNASAR, M., GILLOT, J., GARCERA, D. and TARODO DE LA FUENTE, B. Clarification et stabilisation des vins par ultrafiltration tangentielle sur membranes minérales. *Ind. Aliment. Agric.*, 1984, **101**, 481.

32. GALAJ, S., WICKER, A., DUMAS, J. P., GILLOT, J. and GARCERA, D. Microfiltration tangentielle avec décolmatage des membranes céramiques. *Le Lait*, 1984, **64**, 129.

33. ESCUDIER, J. L., MOUTOUNET, M. and BENARD, P. Influence de l'ultrafiltration sur la cinétique de cristallisation du bitartrate de potassium des vins. *Rev. Fr. Oenol.*, 1987, **27**, 52.

34. GUIZARD, C., ALARY, J. A., LARBOT, A., COT, L., RUMEAU, M., CASTELAS, B. and GILLOT, J. Nouvelles membranes minérales d'ultrafiltration: application à la purification de l'eau. *Le Lait*, 1984, **64**, 276.

35. GAILLARD, M. and BERGER, J. L. Ultrafiltration et microfiltration tangentielle. Résultats d'essais sur moût et sur vin, *Rev. Fr. Oenol.*, 1984, **24**, 39.

36. BARILLERE, J. M., ESCUDIER, J. L., MOUTOUNET, M. and BENARD, P. Quelques applications de la microfiltration tangentielle en oenologie. *Rev. Fr. Oenol.* 1985, **25**, 9.

37. FINNIGAN, T., SHACKLETON, R. and SKUDDER, P. Filtration of beer and recovery of extract from brewery tank bottoms using ceramic microfiltration. Filtec Europa September 22–24, 1987, Utrecht, Netherlands.

38. LOZANO, Y., HEICH, O., BENNASAR, M. and TARODO DE LA FUENTE, B. Transformation des écarts de triage de Kiwifruits par microfiltration sur membrane minérale. *Ind. Aliment. Agric.*, 1986, **103**, 1139.

39. VIEIRA, S. D. A., GOUDEDRANCHE, H., DUCRUET, P. and MAUBOIS, J. L. Eléments de fabrication d'un nouveau fromage brésilien de type Minos frescal par le procédé M.M.V., *Tech. Lait*, 1983, **978**, 17.

40. GOUDEDRANCHE, H., MAUBOIS, J. L., DUCRUET, P. and MAHAUT, M. Utilization of the new mineral UF Membranes for making semi-hard cheeses. *Desalination*, 1980, **35**, 243.

41. MAHAUT, M., KOROLCZUK, J., PANNETIER, R. and MAUBOIS, J. L. Eléments de fabrication de fromage de type pâte molle de lait de chèvre à caractère lactique par ultrafiltration de lait acidifié et coagulé. *Techn. Lait. Mark.*, 1986, **1011**, 24.

42. BENNASAR, M., ROULEAU, D., MAYER, R. and TARODO DE LA FUENTE, B. Ultrafiltration of milk on mineral membranes: imposed performance. *J. Soc. Dairy Techn.*, 1982, **35**, 43.

43. PIOT, M., VACHOT, J. C., VEAUX, M., MAUBOIS, J. L. and BRINKMAN, G. E. Ecrémage et épuration bactérienne du lait entier cru par microfiltration sur membrane en flux tangentiel. Techn. Lait. Mark. 1987, **1016**, 42.
44. PIOT, M., MAUBOIS, J. L., SCHAEGIS, P., VEYRE, R. and LUCCIONI, L. Microfiltration en flux tangentiel des lactosérums de fromagerie. Le Lait, 1984, **64**, 102.
45. TADDEI, C., AIMAR, P., DAUFIN, G. and SANCHEZ, V. Etude du transfert de matières lors de l'ultrafiltration de lactosérum doux sur membrane minérale. Le Lait, 1986, **66**, 371.
46. OZAWA, K., KIM, H. B., SAKURAI, H., TAKESAWA, S. and SAKAI, K. Novel utilization of ceramic membranes in plasma treatment. In Progress in artificial organs (ed. by Nose, Kjellstrand and Ivanovich), ISAO Press, Cleveland, USA, 1986.
47. GERSTER, D. and VEYRE, R. Mineral ultrafiltration membranes in industry. In Reverse osmosis and ultrafiltration (ed. by S. Sourirajan and T. Matsuura), ACS, Washington, USA, 1985, p. 225.

Chapter 5

PROCESS EVALUATION IN ASEPTIC PROCESSING*

SUDHIR K. SASTRY

Department of Agricultural Engineering, The Ohio State University, USA

SUMMARY

The technology of aseptic processing of food, both liquid and solid–liquid mixtures, is currently of great industry interest. The problem, particularly with large particulate matter, is the assurance of microbiologically safe product. This chapter contains a discussion of mathematical and microbiological methods for evaluation of thermal process schedules, primarily aimed at particulate foods, although some information pertinent to liquid foods is also presented. Public domain mathematical models are evaluated for scraped surface heat exchange systems and potential approaches are described for the ohmic heating technology. Solid–liquid flow and interfacial heat transfer phenomena are highly important in continuous flow systems, and the state of the literature on suspension flows is discussed. Finally, some of the available microbiological approaches for process evaluation are presented.

*Although the recommendations in this chapter are based on scientific studies, references to operating procedures and methods or types of instruments and equipment are not to be construed as a guarantee that they are sufficient to prevent damage, spoilage, loss, accidents or injuries resulting from use of this information. The study and use of this chapter by any person or company is not sufficient assurance that said person or company is proficient in the operations and procedures discussed in this chapter. The use of the statements, recommendations, or suggestions contained herein is not to be considered as creating any responsibility for damage, spoilage, loss, accident or injury resulting from such use.

NOTATION

A	Cross sectional area (m^2)
C_p	Specific heat (J/kg°C)
CCF	Convective coefficient factor, the ratio of fluid-to-particle convective coefficient assumed, to the wall-to-fluid value from the literature
F_o	Lethality (min at 121°C)
h_c	Convective heat transfer coefficient (W/m^2°C)
HTRTR	Holding tube residence time ratio, the ratio of residence time for the specific particle, to the mean residence time for the product within the holding tube
HXRTR	Heat exchanger residence time ratio, the ratio of residence time for the specific particle, to the mean residence time for the product within the heat exchanger
I	Current (amp)
J	Current density (amp/m^2)
k	Thermal conductivity (W/m°C)
L	Length (m)
m	Temperature coefficient of electrical conductivity (°C^{-1})
n	Flow behavior index (dimensionless)
n	Unit normal vector
q'''	Rate of energy generation per unit volume within liquid medium (W/m^3)
q_p'''	Rate of energy generation per unit volume within particle (W/m^3)
Q	Rate of energy generation or dissipation (w)
R	Electrical resistance (ohms)
SR	Size ratio, or dimension of particle under consideration to dimension of average sized particle
t	Time (s, or min for lethal values)
T	Temperature (°C)
v	Fluid velocity (m/s)
V	Volume (m^3) or fluid velocity (m/s)
Z	Z-value (time required to achieve one log decrease in thermal death time, min)
α	Thermal diffusivity (m^2/s)
∇	Gradient
ρ	Density (kg/m^3)

σ	Electrical conductivity (ohm m)$^{-1}$
∂	Operator

Subscripts

max	Maximum
mean	Mean value
s	Surface
T	At temperature T
x	Coordinate direction
y	Coordinate direction
z	Coordinate direction
25	At 25°C
∞	Bulk value for fluid

1. INTRODUCTION AND SCOPE

Aseptic processing, or more precisely, continuous sterilization of foods that are to be aseptically packaged, is a subject of great current interest to food processing industries in many countries. The combination of packaging cost reduction, increased convenience, and the use of ultra high temperature (UHT) processes have created products that have proved desirable for consumers and profitable for processors. This technology has gained widespread international acceptance for liquid foods, notably fruit juices, sauces, puddings and milk (not in the United States).

The success of continuous sterilization for liquids has led to widespread interest in extension of this technology to foods (both acid and low-acid) containing discrete particulate matter. Considerable research is presently ongoing in industry, academic and research institutions in an effort to bring this development to fruition. Among the technical hurdles to be overcome have been the development of pumps to minimize damage to particles, and fillers for accurate metering of suspensions into packages. Systems have been, and are being, developed to meet these challenges.

Perhaps the single most important consideration in the commercial production of aseptically packaged foods (particularly low-acid foods), liquid or particulate suspension, is the assurance of microbiological safety of the product. For this purpose, it is necessary to ensure that all parts of the food product receive sufficient thermal processing to ensure commercial sterility. If the food is liquid, this is relatively simple (in principle), since temperatures of flowing liquid streams can be measured relatively easily.

However, continuous sterilization of low-acid particulate suspensions requires consideration of the possibility that *Clostridium botulinum* spores may be present at the slowest heating locations of individual food particles. Since no reliable technique presently exists for monitoring temperatures within cold zones of individual free-flowing particles, it is necessary to estimate the cold-zone temperature history using a suitable mathematical model, and then verify the model by microbiological experimentation.

This chapter is devoted to a review and discussion of some of the public domain methods, mathematical and microbiological, for evaluation of thermal processes for continuous sterilization of foods. Also included in this chapter is a discussion of some of the fundamental flow phenomena influencing thermal process evaluation. It is not intended to be a 'do-it-yourself' manual for those desiring to evaluate aseptic processes for their products, but rather as a reference that may be helpful in identifying important parameters, so that the processor may understand some of the considerations used by the process authority. Indeed, the author considers consultation with a process specialist as an essential step in process establishment.

It is recognized that methods of process evaluation other than those discussed for particulate foods exist in private industry, but these are not widely available and cannot be included in the present chapter. The major focus of this chapter is process evaluation for particulate suspensions, although some discussion will be presented relative to liquid processes in indirect contact systems.

Since the criteria for commercial sterility are different depending on the product and country of production or sale, the exact process to be delivered to the product is a function of the above parameters. There also exist variations in size and characteristics of process equipment dependent upon the manufacturer, which would have an impact on the process specification. Finally, novel process equipment is constantly under development, (e.g. the ohmic heater developed by the Electricity Council Research Centre in Great Britain, and licensed to APV International) and each new device may require different approaches to thermal process evaluation. Consequently, this chapter will focus primarily on fundamental principles common to systems presently in commercial use, with discussion of specific variations that may occur due to novel process equipment. Unless otherwise specified, all discussion will pertain to a typical heat-hold-cool system such as illustrated in Fig. 1, consisting of heat exchangers for product heating, holding tube for achievement of commercial sterility, and a cooling section consisting of heat exchangers. It will be assumed that steam and water are

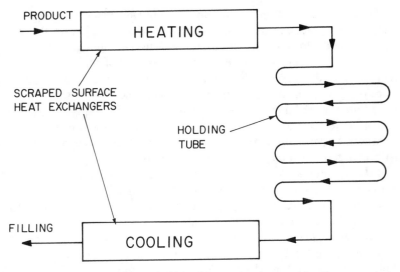

FIG. 1. Typical heat-hold-cool system under consideration.

used as the heating and cooling media respectively. The actual heat exchanger design would vary depending on the product and application, being of tubular or plate type for liquid foods and scraped surface (or wide tube) for particulate suspensions.

2. MATHEMATICAL METHODS

2.1. Liquids

The liquid is heated to process temperature within the heating section. The holding time should be based on the total lethality accumulated within the holding section, given by the equation:

$$F_o = t\,10^{\{(T-121\cdot1)/Z\}} \qquad (1)$$

The total accumulated lethality for any given temperature history may be estimated by numerical integration of eqn (1) over the process time. In the present application, holding time (t) may be calculated algebraically from eqn (1), by using a suitably conservative target value for F_o, and by assuming that T is equal to the temperature measured at the outlet of the holding tube. The temperature measurement should be made at the center of the tube, since fluids in laminar flow are not well mixed, and significant temperature gradients may exist between the tube axis and the wall. A recent model for heat transfer to non-Newtonian fluids is by Pereira et al.[1] If centerline temperatures are measured, the use of holding tube outlet temperature for process calculation represents a conservative approach since the fluid cools during its passage through the holding tube, and the temperature would therefore be lowest at exit. Note that eqn (1) is generally used to determine lethality at temperatures common to can processing. However, evidence does exist that indicates that thermal death character- istics of organisms could be different from the above assumption at the ultra high temperatures common to aseptic processing. Burton et al.[2] found considerable discrepancies between data obtained (for *Bacillus stearothermo- philus* in milk) from capillary tubes and UHT sterilizers, and urged caution in process evaluation. Srimani et al.[3] found that the D-value for *B. stearothermophilus* (without heat activation) and *B. subtilis* remained constant above 135°C and 120°C respectively. Some researchers including Manji and van de Voort[4] favor the Arrhenius model over the Thermal Death Time (TDT) model for temperature dependence of bacterial spore death, although recognizing that the differences may not be of practical significance. Rodriguez et al.[5] have developed a sophisticated model for spore death kinetics based on considerations of population dynamics and systems analysis. These and other studies suggest that the process designer must use caution in extrapolation of TDT data to aseptic temperatures, and a suitable margin of safety should be employed in each situation.

The minimum safe holding tube length must provide sufficient residence time to render all parts of the fluid commercially sterile. Since the fluid at the center of the tube moves with the maximum velocity, possessing the shortest residence time at process temperatures, the calculation of holding tube length must be based on ensuring sterility of this fluid component. The maximum fluid velocity (V_{max}) in tube flow is related to the mean fluid velocity (V_{mean}), the exact nature of the relationship depending on the rheological properties of the fluid and the flow regime (laminar, transitional or turbulent) under consideration. For a Newtonian fluid in laminar, fully developed flow under a constant pressure gradient, the solution to the Navier–Stokes equations in one dimension yields:

$$V_{max} = 2V_{mean} \tag{2}$$

For a Newtonian fluid in turbulent flow, the relationship is:[6]

$$V_{max} = 1\cdot2V_{mean} \tag{3}$$

For a power law fluid (with flow behavior index n) in laminar flow, the solution to the fluid momentum equations (subject to the same assumptions as the Newtonian laminar flow case) yields:

$$V_{max} = \{(3n+1)/(n+1)\} V_{mean} \tag{4}$$

In the limiting case of a Newtonian fluid ($n=1$), eqn (4) reduces to eqn (2). For a pseudoplastic fluid ($n<1$), $V_{max} < 2V_{mean}$, while for a dilatant fluid ($n>1$), $V_{max} > 2V_{mean}$. Turbulent flow relationships between V_{max} and V_{mean} for non-Newtonian fluids are not as well defined as for Newtonian fluids, and experimental data on residence time distributions is necessary for design situations. The conservative approach, of course is the assumption of laminar flow.

The minimum holding tube length may then be calculated as:

$$L = t V_{max} \tag{5}$$

where t is determined from the kinetic considerations discussed above, and V_{max} from V_{mean} as described above, knowing the product flow rate, pipe cross sectional area, flow regime and fluid rheological behavior. It is, of course, desirable to test the calculated process by microbiological procedures, as performed by process specialists. Some useful references in this connection include Bernard[7] and CFPRA.[8]

The preceding discussion does not include consideration of residence time distributions in steam injection or steam infusion systems. Such a study has been published by Heppell[9] for a steam infusion sterilizer, using water and milk as test fluids. He found that milk exhibited a wider residence time distribution than water. Such studies are necessary for each fluid being considered for aseptic processing; however, further discussion is not within the scope of this article.

2.2. Particulates Suspended in Liquid

Process evaluation for large food particle suspensions is considerably more difficult than for liquids. The principal difficulty is in measurement of cold-zone temperatures of individual food particles during continuous flow. When a suspension is heated within heat exchangers, the liquid heats up relatively rapidly, but particle thermal response may be considerably slower. Further complication is introduced by the existence of residence time distributions in process equipment (both heat exchanger and holding tubes) and the unknown value of fluid-to-particle convective heat transfer coefficient. Consequently, process evaluation depends on development of accurate mathematical models for prediction of particle cold-spot temperatures, and reliable techniques for microbiological validation.

Mathematical models for thermal process evaluation for particulates consider the response of a representative food particle to changing environmental temperature. A particle in a continuous sterilization stream is exposed to liquid undergoing changing temperature. Consequently, the temperature distribution within the food particle is given by the solution to:

$$\nabla \cdot (k \nabla T) = \rho C_p (\partial T/\partial t) \tag{6}$$

For constant thermal conductivity (typical for most food particles in thermal processing), eqn (6) may be rewritten in scalar three-dimensional form (in Cartesian coordinates) as:

$$\frac{\partial^2 T}{\partial x^2} + \frac{\partial^2 T}{\partial y^2} + \frac{\partial^2 T}{\partial z^2} = \frac{1}{\alpha} \frac{\partial T}{\partial t} \tag{7}$$

with the time-dependent convective boundary condition:

$$k \nabla T \cdot \mathbf{n} = h_c (T_s - T_\infty(t)) \tag{8}$$

which can be written in scalar form only for regular shaped objects. Since many food particles are of irregular shape, the boundary condition above will be left in vector form, which can be easily addressed during numerical solution.

2.2.1. Fluid Medium Temperature

Solutions to eqn (7) subject to boundary condition eqn (8) requires that the fluid medium temperature T_∞ be known as a function of time or of location within the system (since the location is generally expected to vary linearly with time, the two variables are interchangeable with a minor transformation). The easiest approach is direct measurement of liquid medium temperatures at various locations within the system, and by using the assumption that the fluid is well mixed. Another approach is also possible; by conducting energy balances on fluid and particles, calculating the fluid medium temperature with the well mixed fluid assumption. In both approaches, it is desirable to obtain a suitably conservative estimate of fluid medium temperature to account for gradients that may actually exist in the fluid. In general, (with both procedures) the field temperature may be expected to follow the exponential forms illustrated in Fig. 2.

2.2.2. Particle Temperature Distributions

Solution of eqn (7) subject to boundary condition eqn (8) has been attempted (with various simplifications) by some researchers. One of the earliest attempts was by de Ruyter and Brunet[10] who obtained an analytical solution, using the following assumptions.

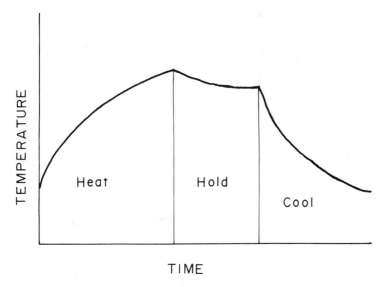

FIG. 2. Representative exponential fluid temperature change patterns for various sections of the system.

1. Particles are spherical.
2. Particles are isotropic with regard to thermal and physical properties.
3. Total particle mass is not greater than 25% of the whole system (an arbitrarily chosen figure below which two distinct phases could be assumed).
4. It is permissible to use experimentally determined overall heat transfer coefficients to compensate for heat changes of the fluid phase due to changes of temperature of the solid particles.
5. Initial temperature of solid and liquid phases are equal and uniform.
6. Linear (ramp function) temperature changes occur in the liquid phase during heating, holding, cooling and discharge segments.
7. Product is well mixed.
8. Particle surface temperature is equal to the fluid medium temperature, implying an infinite fluid-to-particle convective heat transfer coefficient.

Assumptions 4, 6, and 8 have the most significant effects on process design. The meaning of assumption 4 as stated by the authors is unclear, but the text of the paper suggests that the temperature of the liquid phase was calculated from a lumped energy balance, equating the heat transferred across the walls to the heat gained by the liquid. This energy balance does not include the consideration that the particles act as thermal 'drags' on the liquid; however the mass flow rate of liquid appears to be set equal to the overall product flow rate (it is not explicitly stated by the authors). If this is indeed the case, the assumption is made that the liquid–particle mixture could be treated as an overall 'fluid' thereby increasing the mass flow rate and decreasing the rate of rise or drop of predicted liquid temperature. This approach would lead to errors in estimation of liquid temperature, the exact nature of which would depend on the specific situation. It is unclear how the authors have compensated for the error due to particle thermal drag by choice of the overall heat transfer coefficient.

Assumption 6 (ramp function changes in liquid temperature) may result in error due to deviation from true liquid medium temperatures. Whether or not these errors are conservative or significant depends on the particular product, equipment zone and processing situation. If the linear liquid temperature approximations are as shown in Fig. 3 (a similar figure is used by the authors), the expected pattern would be a low (and conservative) temperature prediction during heating, and a high (and non-conservative) temperature prediction during holding and cooling.

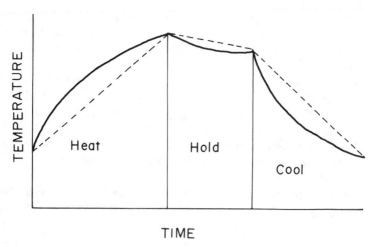

FIG. 3. Comparison of exponential and linear temperature profiles for various sections of the system.

Assumption 8 is perhaps the most critical, since the assumption of infinite fluid-to-particle convective coefficient would result in substantial over-prediction of particle cold spot temperatures within heating and holding segments. Though the model would tend to underpredict particle cold-spot temperatures during cooling, this cannot be reliably used as a margin of safety because of the potential for particle disintegration and accelerated cooling in this zone. The result of this assumption, therefore, is potential undersizing of the processing system.

The other assumptions in de Ruyter and Brunet's work are somewhat less significant with regards to process evaluation, and are mainly restrictive in character. The assumption of spherical shape (and symmetry) reduces the governing differential equation to a one-dimensional transient problem which is relatively easy to solve, although it tends to restrict the applicability of the model. Another feature of the work is the absence of any mention of residence time distributions within the processing system. Since this is a phenomenon of considerable significance, its occurrence should be accounted for in some manner. A knowledgeable user could conceivably be able to study this effect by changing time scales within the de Ruyter and Brunet model. Despite the assumptions and limitations noted above, the de Ruyter and Brunet work is notable in that it represents some of the earliest work on the subject and sheds light on some of the important phenomena involved. Additionally, the authors have recognized that fluid-to-particle boundary layer characteristics in real systems may be different from those actually assumed.

Another important contribution to the literature is by Manson and Cullen,[11] who presented a finite difference solution to eqns (6) or (7) using the approach of Teixeira et al.[12] These researchers demonstrated the importance of consideration of residence time distributions in process equipment. Their analysis was subject to the following major assumptions.

1. Particles are of finite cylindrical shapes.
2. Particle surface temperatures are equal to fluid medium temperatures, thus fluid-to-particle convective coefficients are assumed to be infinite.
3. Liquid temperature rises linearly in the heating section.
4. Two situations are considered within the holding section; (a) constant liquid temperature; and (b) exponential decrease in liquid temperature.
5. Cooling section contribution was not included.
6. Residence time distributions considered within the scraped surface heat exchanger and holding tube.

With the exception of assumption 2, the approach of Manson and Cullen is conservative in practically all respects. The assumption of infinite convective coefficient would result in overprediction of lethality within the heating and holding sections; a point recognized by the researchers. Assumption 3 is conservative, as discussed earlier in connection with the de Ruyter and Brunet model. Assumption 4(a) is conservative if the constant temperature considered is that at the holding tube outlet (as mentioned by the authors), while assumption 4(b) is reasonable if liquid temperatures within the holding tube are accurately known. The authors have rightly pointed out that particles at high temperatures are fragile, and could conceivably disintegrate within the cooling section; thus cooling would be more rapid than anticipated, with the potential for overprediction of cooling section lethality. Consequently, use of cool-down lethalities as safety factors may not be reliable. The authors have used the (scraped surface heat exchanger) residence time distribution information available in the literature at the time (Chen and Zahradnik,[13] for liquids), and used a power-law fluid model to describe the liquid velocity profile within the holding tube. Within the holding tube, particles have been assumed to be in greater concentration towards the tube axis. Residence time distribution effects on integrated lethality have been calculated by considering eleven concentric shells within the holding tube section, and assuming the particles to flow along the fastest moving streamline within each shell. The authors demonstrate the importance of consideration of particle size and residence time distribution effects. The principal limitations of the model are with regard to assumptions 1 (cylindrical shape) and 2 (infinite fluid-to-particle convective coefficient).

Dail[14] has used the Ball formula method for thermal process evaluation within the holding tube only. While the Ball method is relatively simple, its applicability is severely restricted because of the numerous assumptions involved at arriving at the simple result. A discussion of the derivation and limiting assumptions of this method have been provided by Merson et al.[15] Some of the key assumptions and limitations of this method of process evaluation are as follows.

1. Particles are of regular shapes.
2. Fluid-to-particle convective coefficient is infinite.
3. Long process times are necessary for use of the first-term approximation involved in the Ball method.
4. Applicability of the method is subject to convergence of the infinite series solution.

Perhaps the major restriction of this method is that the development of

Ball's formula requires a constant temperature heating medium; consequently the method cannot be used for predicting particle temperatures within the heating section. The method is therefore limited to certain situations within the holding tube only. The restriction of long process times further restricts the analysis to certain food particles only. A further restriction stated by the author is the requirement that there be a difference of at least 80°F between the particle cold zone and the fluid at the entrance to the holding tube. Since this does not occur in many cases, the suggested approach is the assumption that the particle cold-zone temperature at the holding tube inlet is equal to the product initial temperature before heating, thus providing a conservative assumption. The assumption of infinite fluid-to-particle convective coefficient would result in overprediction of particle temperature. However the author proposes to compensate for this deficiency by determining an 'apparent thermal diffusivity' for particles of sizes larger than those to be processed. The exact size relationships necessary for accurate compensation is not specified, but would be expected to depend on the characteristics of the heating medium. While it is possible in principle to correct for the assumption of infinite heat transfer coefficient, the correction can only be considered arbitrary for lack of information. The severely restrictive nature of this method combined with the number of unspecified correction procedures would argue against its widespread use for aseptic thermal process evaluation. This point has been well recognized by the author, who describes the restrictions involved in the approach.

Sastry[16] solved eqn (6) subject to boundary condition eqn (8) using the Galerkin–Crank–Nicolson algorithm involving three-dimensional finite elements in space and finite differences in time. The key assumptions were as follows.

1. Particles are of either regular or irregular shapes (the solution was presented for mushroom-shaped particles).
2. Fluid-to-particle convective heat transfer coefficients are finite, and may be determined for worst-case scenarios as desired. A Nusselt number of 2 was chosen as corresponding to conduction from a still liquid (Geankoplis[17] for a sphere).
3. Fluid medium temperatures determined using incremental energy balances over sections of heat exchanger and holding tube sections.
4. Fluid medium considered to be well mixed in the radial direction of the heat exchanger and holding tube, and at all times in contact with a population of particles of uniform size moving at the same average velocity.
5. Viscous dissipation in scraped surface heat exchanger is negligible.

The assumptions involved in this approach are conservative in nature. Assumptions 3 and 4 are used to calculate fluid medium temperatures and the computational scheme is designed to ensure conservatism in calculation. Simulations for individual food particles are conducted using conservative estimates of fluid medium temperature. The well-mixed fluid assumption may not be truly valid, and is made primarily for convenience in calculation. In reality, the fluid within an incremental section within the scraped surface heat exchanger is in contact with a population of particles of various sizes and ages. The relative influence of particle size on fluid temperature is slight (small particles produce greater thermal drag on the fluid); nevertheless the most conservative fluid temperature estimates available are used. It is assumed that this factor, combined with the conservatism of the computational scheme and the neglect of viscous

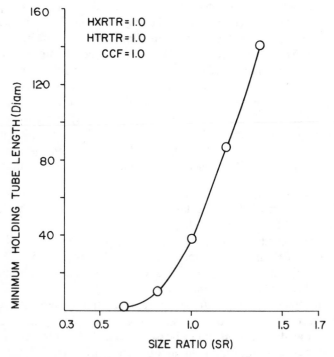

FIG. 4. Effect of dimensionless particle size (size ratio) on minimum required holding tube length (adapted from Sastry,[16] with permission from Institute of Food Technologists).

dissipation combine to give suitably conservative estimates of fluid medium temperature.

Simulations of Sastry[16] reveal the influence of various dimensionless parameters on holding tube sizing (definitions of each parameter are presented in the notation list). In these studies, only lethality accumulated within the holding tube is considered, and that in the heat exchanger ignored. While the exact numerical results are specific for the product, equipment type and simulation, the general trend can be inferred from these studies. The dramatic influence of particle size is illustrated in Fig. 4, showing the increase in dimensionless holding tube length necessary to process large particles. Influence of heat exchanger residence time distribution is also highly significant, as illustrated in Fig. 5. The influences of

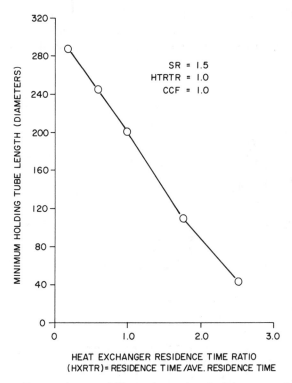

FIG. 5. Effect of heat exchanger residence time ratio on minimum required holding tube length (adapted from Sastry,[16] with permission from Institute of Food Technologists).

fluid-to-particle convective coefficient and holding tube residence time are illustrated in Figs 6 and 7 respectively. These figures represent successively worst case scenarios. These results indicate that the sizing of the holding tube should be based on a hypothetical particle that is the slowest heating in the product by virtue of largest size (or thermal properties conducive to slow heating), most rapid movement through heat exchanger and holding tube, and lowest convective coefficient. It is uncertain whether or not such a particle exists, particularly since fast moving particles would be expected to possess appreciable velocity relative to the fluid, and consequently a relatively high convective heat transfer coefficient. However for lack of reliable information, the only safe assumption is the conservative one, and the above mentioned criteria must be considered until such data are available.

The results of typical simulations are presented in Fig. 8 (reproduced from Sastry,[16] courtesy Institute of Food Technologists), where the temperature history of fluid and particle cold spot are presented for two situations: one involving a suspension of relatively small particles ($SR =$

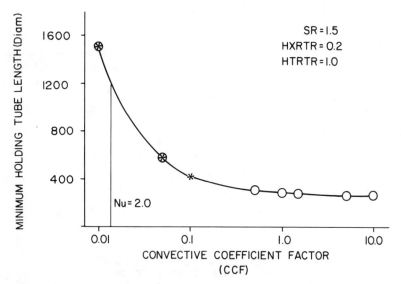

FIG. 6. Effect of dimensionless fluid-to-particle convective coefficient (convective coefficient factor) on minimum required holding tube length (adapted from Sastry,[16] with permission from Institute of Food Technologists).

0·63), and the other, large particles ($SR = 1·37$). For small particles, thermal lags are relatively small, but for the large particles, lags at holding tube inlet could be considerable. Similar general trends have been observed in simulations by Hegg and Belfrage.[18] Simulations by Sastry[16] also show that for small particles, considerable lethality is accumulated within the heat exchanger, while for large particles, heat exchanger contribution to lethality is negligible.

Other modeling studies have been reported in the literature; these include studies by Newman and Steele,[19] and several reports in the IUFoST Symposium on Aseptic Processing and Packaging of Foods.[20] However, detailed discussion of all methods is beyond the scope of this chapter.

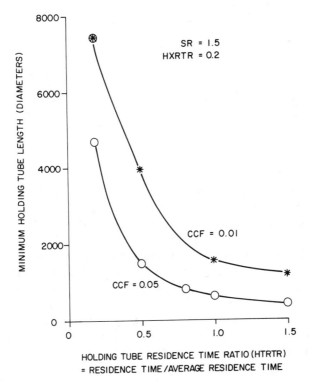

FIG. 7. Effect of holding tube residence time ratio on minimum required holding tube length (adapted from Sastry,[16] with permission from Institute of Food Technologists).

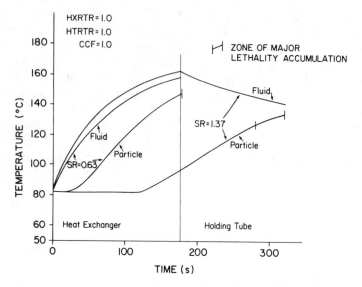

FIG. 8. Fluid and particle cold-zone temperatures within heat and hold sections as influenced by two particle sizes. (Reprinted from *Journal of Food Science*, 1986, **51**(5), 1323, Copyright © by Institute of Food Technologists.)

2.2.3. Ohmic Heating

A recent development of the Electricity Council Research Centre in Great Britain is the use of the product's own electrical resistance for heating. This heater, termed the 'ohmic' heater has been licensed to APV International Ltd, and is under development for use in continuous sterilization of products including liquid foods and foods containing discrete particulates. The principal advantages claimed for the ohmic heater are (1) relatively uniform heating due to internal energy generation within the product rather than by conventional surface heating; (2) gentle handling of particulates; (3) minimal fouling of heat transfer surfaces; and (4) silent operation. The basic principle behind the heater is the I^2R energy dissipation within the product; thus

$$Q = I^2R \tag{9}$$

within a continuous conductor. The resistance offered by the product to flow of electrical current is dependent on the electrical conductivity, as given by the relationship:

$$R = L/A\sigma \tag{10}$$

Since the effectiveness of this method of heating depends on the product's electrical conductivity, it may not be applicable to all food materials. Biss *et al.*[21] have presented a discussion of ohmic heating, in which they state that most pumpable foods with water content above 30% and dissolved ionic salts have potential for ohmic treatment. Non-ionic materials such as fats, oils, or sugar have been found to be unsuited to this heating method.

Electrical conductivity has been found to increase with temperature, following a relationship modeled by Biss *et al.*:[21]

$$\sigma_T = \sigma_{25}\{1 + m(T - 25)\} \tag{11}$$

Process evaluation for ohmic heating represents a somewhat different problem from conventional heating. For a pure liquid, the energy equation may be written as:

$$\rho C_p(DT/Dt) = \nabla \cdot (k\nabla T) + q''' \tag{12}$$

where the derivative on the left hand side is the substantial derivative, defined as:

$$\frac{DT}{Dt} = \frac{\partial T}{\partial t} + v_x\frac{\partial T}{\partial x} + v_y\frac{\partial T}{\partial y} + v_z\frac{\partial T}{\partial z}$$

For a well-mixed fluid, the equation reduces to:

$$\rho C_p(dT/dt) = q''' \tag{13}$$

where

$$q''' = Q/V = I^2R/V \tag{14}$$

Note that q''' is a temperature dependent quantity. Using the common definitions of current density and the relationship for electrical resistance (eqn (10)) Biss *et al.*[21] have presented the following equation for heating of well mixed liquids in the ohmic heater:

$$dT/dt = J^2/\sigma_T\rho C_p \tag{15}$$

For foods containing particulates, the problem of process evaluation involves the solution to the energy equation for conduction heat transfer into particulate foods:

$$\nabla \cdot (k\nabla T) = \rho C_p(\partial T/\partial t) + q'''_p \tag{16}$$

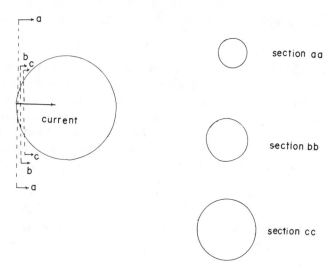

FIG. 9. Change in cross section exposed by food particle with depth of penetration
of electrical current.

subject to the same time-dependent convective boundary condition
represented by eqn (8). The energy generation rate per unit volume within
the particle (q_p''') would be expected to depend on location within the
particle as well as temperature. Since the cross section presented by the
particle to the heating current would change with depth of penetration (see
Fig. 9), the energy dissipation at different locations of the particle would be
expected to vary as well. If the particle is of irregular shape, rotation of the
particle would cause a set of time-varying cross sections to be exposed to the
heating current; consequently q_p''' could be a function of time, temperature
and spatial location within each particle. For composite particles, or
particles of non-uniform composition, the problem is even more complex.
Further, these considerations do not include residence time distributions;
consequently numerous challenges remain to be met. If the fluid is a good
conductor, but the particle is not, it is possible for the current to bypass the
particles entirely, reducing the problem once again to one of pure
conduction without energy generation.

3. SIGNIFICANT FLOW PHENOMENA

While a number of approaches may be taken towards modeling of conduction heat transfer in particulate suspensions, real data on residence time distributions and fluid-to-particle convective coefficients are scarce. Indeed, lack of data represents one of the major roadblocks to development of aseptic processing technology for particulates. This represents an immediate research need. Although results would likely be product-specific, good contributions would be particularly helpful. Both residence time distributions and convective heat transfer coefficients fall under the broader category of flow phenomena, and are likely to be interdependent. These phenomena will be discussed in this section.

Relatively few studies have been conducted to determine fluid-to-particle convective coefficients, an understandable situation in light of the difficulties of particle temperature measurement in continuous flow. Heppell[22] attempted to measure convective coefficients in continuous flow systems by using alginate beads to immobilize spores of *B. stearothermophilus* within 3·1 mm diameter alginate beads, using the approach of Dallyn *et al.*[23] He then subjected the beads to a continuous sterilization process, measured spore survivor concentrations, and by fitting these data to a mathematical model, calculated the convective heat transfer coefficient. It was found that at a Reynolds number of 50 000, the average convective heat transfer coefficient was 7300 W/m^2°K, corresponding to a Biot number of 18·6, but at a Reynolds number of 5300, the value was 1850 W/m^2°K corresponding to a Biot number of 4·3. The author concluded that the marked reduction in heat transfer coefficient with decreasing Reynolds number was due to decreased turbulence. Although this method represents an ingenious approach to a difficult problem, the convective heat transfer coefficient values need verification, and seem to be somewhat too high for use as process design assumptions.

Zuritz *et al.*[24] attempted to estimate fluid-to-particle heat transfer coefficients in a tube flow situation by using a mushroom-shaped metal transducer particle immobilized within a tube, and pumping a non-Newtonian fluid through the tube. They obtained dimensionless correlations for the Nusselt number as a function of generalized Reynolds and Prandtl numbers. Convective heat transfer coefficients were found to range from 548–1175 W/m^2°K for the range of fluid and particle variables studied. Convective coefficients were found to increase with increasing particle size and fluid flow rate, and with decreasing consistency coefficient. The trend with particle size is readily explainable. Since the particles were immobilized, increasing particle size would decrease the cross section available to the fluid for tube flow. The fluid would consequently be accelerated past the particle resulting in an experimental artifact. The authors urge caution in use of the data since they represent upper limit values for the case of static particles during tube flow and do not represent actual processing conditions.

Residence time distribution has received somewhat greater attention in the literature than convective coefficient, although most studies[9,13,25,43] apply to liquid foods only. Taeymans et al.[26] studied residence time distribution of solid–liquid mixtures during continuous sterilization in a scraped surface heat exchanger, using a suspension of 6 mm diameter calcium alginate beads in water. They observed that dasher speed had significant effects on the residence time, and that the mean residence time for the beads was greater than the calculated mean residence time of fluid (ratio of heat exchanger volume to liquid flow rate). No other variables (such as fluid rheological properties or particle size) were investigated.

McCoy et al.[27] investigated single-particle behavior in a viscous non-Newtonian fluid within a simulated transparent holding tube (inclined upwards at an angle of 1·194° as required by the US Food and Drug Administration) by introducing plastic particles (densities ranging from 1003–1053 kg/m^3) of various sizes into flowing solutions of sodium carboxymethylcellulose (CMC). Variables studied included particle size, fluid medium viscosity, fluid flow rate and the influence of bends. Several qualitative observations of particle motion were also conducted. Some general effects noted were as follows: (1) for highly viscous fluids, bend size had little effect on particle radial position (and consequently, particle velocity); (2) bends had significant effects on particle radial position and velocity when low-viscosity fluids were used; (3) the influence of bends increased with increasing fluid flow rate; (4) the most stable and consistent behavior was observed with large particles; (5) small particle behavior was more unstable and residence times were found to be more widely distributed for these particles; (6) the initial radial position of the particles (at the holding tube entrance) was found to have significant effects on particle residence times (particularly for small particles). These studies are for single particles only, and do not include the effects of particle populations. They cannot therefore be considered representative of actual process conditions; nevertheless the study brings to light some of the basic phenomena influencing particle behavior in tube flow. One of the observations of interest was the tendency for certain particles to migrate towards the tube wall. Upon reaching the wall, particles were observed to exhibit a combination of rolling and sliding motion, which was dependent on fluid viscosity. For high viscosity fluids, motion was primarily by sliding, while for lower viscosity fluids, rolling motion predominated. It was not possible to prevent variations in particle density in these experiments, and the results may have been influenced accordingly. The phenomena observed would appear to result from complex combinations of drag, lift, buoyancy and centrifugal forces. Due to the required inclination of holding tubes, the buoyancy force possesses a component tending to either accelerate or decelerate a particle along the flow path. Centrifugal forces at bends could tend to force a particle towards or away from the tube axis. The complexity of these phenomena suggest a necessity for understanding suspension flow at a fundamental level. Some basic understanding may be gained by studying the literature on solid–liquid flows. Sastry and Zuritz[28] have presented a review of some of the relevant literature.

Much of the literature on solid–liquid flows has concentrated on mineral slurry transport, and is not relevant to the holding tube situation. However, studies on capsule flow (a pipelining concept developed in Canada in which cylindrical capsules of diameter nearly equal to the pipe are pumped in liquid suspension) may possess potential relevance. Studies by Hodgson and Charles[29] indicated that capsules always moved faster than the fluid medium (usually water) since they generally flowed along the axis of the pipe. Smaller capsules moved along the central, rapid moving streamlines, and moved faster than capsules of larger diameter. Further studies conducted by Ellis[30] indicated that the smallest cylindrical capsules tended to follow a wavy, unstable motion or to·migrate radially to the walls of the pipe. Spherical capsules of the smaller diameters were also observed to migrate away from the tube axis, and move at slower velocities than expected.

The radial migration of particles in pipe flow has been the subject of considerable interest in the literature, since it was observed by Segre and Silberberg[31] that neutrally buoyant particles would migrate away from both the tube axis and wall, reaching an equilibrium radial position of 0·6 times the pipe radius from the axis. Expressions have been developed for lift forces causing radial migration by Rubinow and Keller[32] and by Saffman.[33] The expression of Rubinow and Keller applies to spinning particles only, and is the Magnus force associated with curving ball trajectories well known to golfers, tennis players, and other sports participants. The expression of Saffman (a corrected expression is presented by Saffman[34]) is applicable regardless of particle rotation. Both these expressions have been developed for creeping flow situations, where inertial forces are negligible.

Other studies, notably that of Cox and Brenner[35] treat the situation where the particles are close to the wall.

Although the expressions for radial migration forces are useful for understanding of flow phenomena, they have been derived using restrictive assumptions, and may not be directly applicable to real world situations. Also, these studies relate to single particle situations, and do not address the behavior of a population of particles. Studies by Segre and Silberberg indicate that in steady flow, particles that became concentrated in a thin cylindrical layer and all moving at the same velocity aligned themselves into regular columns or 'necklaces'.

Much of the difficulty involved in studies of particle populations is that involved in modeling many-particle hydrodynamic interactions. The computational challenges posed by these problems are formidable, and most researchers have used severely restrictive assumptions to permit analyses. An attempt to understand radial migration in realistic pipe flow situations is by Lawler and Lu[36] who modeled the trajectory and velocity of a spherical particle in vertical pipe flow, by using a combination of expressions for lift forces as given by Saffman[33] and Rubinow and Keller.[32] These authors were apparently unaware of the correction to Saffman's expression presented in 1968.[34] They claimed to be able to obtain qualitative agreement with experimental data by using this approach.

Durst et al.[37] have developed an approach for predicting multiple particle trajectories and velocities in tube flow, using a finite difference solution to the Navier–Stokes and particle dynamics equations. However, the approach neglects particle–particle interactions, and is restricted to small particles. Despite these limitations, it represents one of the best available approaches at present.

Sastry and Zuritz[38] have developed a model to predict particle trajectories and velocities in a holding tube situation (inclined upwards at 1·194°), using a modification of the approach of Lawler and Lu.[36] The model considers radial migration, buoyancy, drag and lift forces, and is usable for straight pipe flow when particles are not in contact with the wall. The model does not presently account for situations involving centrifugal forces (as encountered during flow in bends), or for situations where particles slide and roll in the vicinity of the wall. Since the expressions used are those derived in the literature with restrictive assumptions, the model is only useful for qualitative understanding of flow phenomena.

Flow behavior of particle suspensions is of great importance to future developments in processing of particulate foods. Due to the present lack of information, the process designer can only use the most conservative assumptions possible to ensure a safe product. While this may be workable for some products, it is unlikely to be universally applicable. It is expected that with the present concentration of research activity in this area, better understanding can be expected in the years ahead.

4. MICROBIOLOGICAL EVALUATION

An essential step in aseptic thermal process evaluation is the microbiological verification of mathematical models. Reliable microbiological testing of continuous particulate sterilization necessitates use of a biological indicator that can be processed in a continuous sterilization system and tested for sterility thereafter. The indicator must possess a population of thermally resistant bacterial spores that experience temperature histories similar to those within real food particles. Some of the desired characteristics of a biological indicator are: (1) large size (representative of and preferably larger than the largest particle size to be processed); (2) presence of heat resistant bacterial spores throughout the interior, and especially in the cold zones; (3) geometry and thermal properties similar to those of real food particles; (4) retention of spores without leakage through all process steps; (5) physical durability sufficient to withstand process stresses without disintegration; (6) visual distinguishability from real particles, to facilitate recovery; and (7) shelf stability, permitting use when desired, without elaborate preparation procedures.

Dallyn et al.[23] developed a method for immobilization of bacterial spores in an alginate gel. The procedure consisted of mixing bacterial spore suspension into sterile sodium alginate solution, and dispensing drops of the mixture via a syringe, into a sterile calcium chloride solution, to induce gelation. These beads were reported to be of sufficient mechanical strength to withstand processing in a scraped surface heat exchanger. Unfortunately, the largest bead diameter obtained by this procedure was 3·2 mm, which is too small for use in a system involving large particles. Nevertheless, the experiment represents a pioneering effort in that it has been used by a number of later researchers, including Bean et al.[39] and Heppell.[22]

Brown et al.[40] prepared large (0·8–2·4 cm) biological indicators by immobilizing spore suspensions in cubes consisting of alginate mixed with puréed potatoes, peas or meat. They conducted controlled heating studies on these indicators within an experimental test apparatus, and compared the experimental integrated lethality values with values calculated using a procedure derived from Stumbo.[41] They observed differences between experimental and calculated integrated lethal values and attributed these partly to particle shrinkage, limitations of the procedure for determination of D- and Z-values, and limitations in the prediction procedure. They attempted to make some of their bioindicators visually distinguishable, but did not report any attempts at shelf-stabilization. Despite the lack of agreement between experimental and theoretical values, the approach seems to have potential for further investigation.

Sastry et al.[42] developed a biological indicator by infusing suspensions of B. stearothermophilus in alginate solution with a thermostable dye (for identifiability) into mushrooms; followed by gelation of the alginate to immobilize spores. Samples were then freeze-dried for shelf-stabilization. Samples were tested by infusion tests, thermal response studies, leakage tests and thermal processing procedures. The results of the first three tests showed that (1) spores were infused into particle cold zones; (2) thermal response of particles was similar to real mushrooms; (3) spores were retained within samples through reconstitution and blanching procedures. Results on thermal processing showed good agreement between predicted and calculated spore concentrations for short process durations; however, for long process times, experimental survivor concentrations were considerably higher than predicted. This result was considered to be due to the presence of a native population of thermophiles within mushrooms, whose presence was observed only when the infused spore population count was sufficiently depleted. The results suggest that the indicators have potential for further use. However, they have not been tested for physical durability under process conditions.

Microbiological validation procedures are currently being pursued in industry and academic laboratories. One approach that has been used with meat cubes is the immobilization of spores on a thread, and insertion of the thread through the cube. Other approaches have involved the use of materials other than foods as biological indicators. Development of reliable microbiological validation processes is clearly of great importance, and represents a major research need. However, in dealing with thermal processing of particulates, it is important that the microbiologist and engineer work together in process validation. The biological indicator and the model that it is intended to verify must be compatible. For example, if the product to be processed possesses particles shaped like finite cylinders, the model and biological indicator should be developed accordingly. Since mathematical models are only as accurate as their underlying assumptions permit, it is important to check them for conservatism in process prediction. If the biological tests and mathematical model are ill-matched, the results could be potentially disastrous.

5. CONCLUSIONS AND FUTURE NEEDS

Although aseptic processing technology is at an exciting stage of development, it is important to proceed with caution in extending process methods to particulate foods. Because of the difficulty in temperature measurement, future approaches for process evaluation will likely be considerably more sophisticated than, and bear little resemblance to the methods used for conventional can processing. Process evaluation efforts must be truly interdisciplinary; involving an understanding of heat transfer, multiphase flow and microbiology.

The mathematical models used for process development should be accurate, with any errors being on the conservative side. While the methods for liquids are relatively simple, the same degree of caution is necessary. An attempt has been made in this chapter to review some of the public domain methods available for process evaluation, together with a discussion of the assumptions of each. The user of models would be well advised to check them for assumptions or refer to a qualified consultant or process specialist before proceeding with process evaluation.

Since mathematical models rely upon real data on fluid-to-particle convective coefficients and residence time distributions, it is important to obtain realistic data for each situation. Due to paucity of relevant literature, this would appear to be an important research area of the future. Of equal importance is a fundamental understanding of the behavior of food particle suspensions in continuous flow systems.

Process evaluation cannot be considered complete without reliable microbiological validation procedures. While some literature is available on the subject, further research is necessary for ensuring applicability to a wide range of process situations. It is necessary that validation procedures involve an understanding of the model that is to be validated, and its limiting assumptions.

If the three areas mentioned above are approached carefully, the result could be safe, high quality aseptic products to add to the already expanding market. With the extensive research activity in the area, this may only be a matter of time.

REFERENCES

1. PEREIRA, E. C., BHATTACHARYA, M. and MOREY, R. V. Modeling heat transfer in non-Newtonian liquids. ASAE paper No. 87-6539. American Society of Agricultural Engineers, St. Joseph, MI 49085-9659, USA, 1987.

2. BURTON, H., PERKIN, A. G., DAVIES, F. L. and UNDERWOOD, H. M. Thermal death kinetics of *Bacillus stearothermophilus* spores at ultra-high temperatures. III. Relationship between data from capillary tube experiments and from UHT sterilizers. *J. Food Technol.*, 1977, **12**, 149–61.

3. SRIMANI, B., STAHL, R. and LONCIN, M. Death rates of bacterial spores at high temperatures. *Lebensm. Wiss. u. Technol.*, 1980, **13**, 186–9.

4. MANJI, B. and VAN DE VOORT, F. R. Comparison of two models for process holding time calculations: convection system. *J. Food Prot.*, 1985, **48**, 359–63.

5. RODRIGUEZ, A. C., TEIXEIRA, A. A. and SMERAGE, G. H. Kinetic modeling of heat-induced reactions affecting bacterial spores. Presented at the 47th Ann. Mtg Inst. Food Technol., Las Vegas, NV, June 16–19, 1987.

6. GAVIN, A. Thermal process establishment for low-acid aseptic products containing large particulates. *Proc. Natl Food Proc. Assn Conf. Capitalizing on Aseptic II*, The Food Processors Institute, Washington, DC, 1985, pp. 55–7.

7. BERNARD, D. Microbiological considerations of testing aseptic processing and packaging systems. *Proc. Natl Food Proc. Assn Conf. 'Capitalizing on Aseptic'*, The Food Processors Institute, Washington, DC, 1983, pp. 13–14.

8. CFPRA. *Good manufacturing practice guidelines for the processing and aseptic packaging of low-acid foods. Part 1. Principles of design, installation and commissioning.* Campden Food Pres. Res. Assoc., Technical Manual No. 11, Chipping Campden, UK, 1986.

9. HEPPELL, N. J. Comparison of residence time distributions of water and milk in an experimental UHT sterilizer. *J. Food Eng.*, 1985, **4**, 71–84.

10. DE RUYTER, P. W. and BRUNET, R. Estimation of process conditions for continuous sterilization of foods containing particulates. *Food Technol.*, 1973, **27**, 44–51.

11. MANSON, J. E. and CULLEN, J. F. Thermal process simulation for aseptic processing of foods containing discrete particulate matter. *J. Food Sci.*, 1974, **39**, 1084–9.

12. TEIXEIRA, A. A., DIXON, J. R., ZAHRADNIK, J. W. and ZINSMEISTER, G. E. Computer optimization of nutrient retention in the thermal processing of conduction heated foods. *Food Technol.*, 1969, **23**, 845.

13. CHEN, A. C. Y. and ZAHRADNIK, J. W. Residence time distribution in a swept-surface heat exchanger. *Trans. ASAE*, 1967, **10**, 508–11.

14. DAIL, R. Calculation of required hold time of aseptically processed low-acid foods containing particulates utilizing the Ball method. *J. Food Sci.*, 1985, **50**, 1703–6.

15. MERSON, R. L., SINGH, R. P. and CARROAD, P. M. An evaluation of Ball's formula method of thermal process calculations. *Food Technol.*, 1978, **32**, 66–72, 75.

16. SASTRY, S. K. Mathematical evaluation of process schedules for aseptic processing of low-acid foods containing discrete particulates, *J. Food Sci.*, 1986, **51**, 1323–8, 1332.

17. GEANKOPOLIS, C. J. *Mass Transport Phenomena*, Holt, Rinehart and Winston, New York, USA, 1978.

18. HEGG, P.-O. and BELFRAGE, B. *Continuous sterilization of new food products including particulates.* Unpublished document. Alfa Laval Food and Dairy Engineering AB, Box 64, S-221 00 Lund, Sweden, 1987, 13pp.

19. NEWMAN, M. E. and STEELE, D. A. *Food particle sterilization. Technical memorandum No. 210.* Campden Food Pres. Res. Assoc., Chipping Campden, UK, 1978.
20. IUFoST. *Proceedings–IUFoST Symposium on Aseptic Processing and Packaging of Foods,* Lund Institute of Technology, PO Box 118, S-221 00 Lund, Sweden, 1985, 513 pp.
21. BISS, C. H., COOMBES, S. A. and SKUDDER, P. J. *The development and application of ohmic heating for the continuous heating of particulate foodstuffs.* Unpublished document. APV International Ltd, Crawley, West Sussex, UK, 1987, 10pp.
22. HEPPELL, N. J. Measurement of the liquid–solid heat transfer coefficient during continuous sterilization of liquids containing solids. Presented at the 4th Intl Cong. Engr. Food, Edmonton, Alberta, Canada, July 7–10, 1985.
23. DALLYN, H., FALLOON, W. C. and BEAN, P. G. Method for the immobilization of bacterial spores in alginate gel. *Lab. Pract.,* 1977, **26**, 773–5.
24. ZURITZ, C. A., McCOY, S. C. and SASTRY, S. K. Convective heat transfer coefficients for non-Newtonian flow past food shaped particles. ASAE paper No. 87-6538. American Society of Agricultural Engineers, St. Joseph, MI 49085-9659, USA, 1987.
25. ROIG, S. M., VITALI, A. A., ORTEGA-RODRIGUEZ, E. and RAO, M. A. Residence time distribution in the holding section of a plate heat exchanger. *Lebensm. Wiss. u. Technol.* 1976, **9**, 255–6.
26. TAEYMANS, E., ROELANS, J. and LENGES, J. Residence time distributions in a horizontal SSHE used for UHT processing of liquids containing solids. Presented at the 4th Intl Cong. Engr. Food, Edmonton, Alberta, Canada, July 7–10, 1985.
27. McCOY, S. C., ZURITZ, C. A. and SASTRY, S. K. Residence time distributions of simulated food particles in a holding tube. ASAE paper No. 87-6536. American Society of Agricultural Engineers, St. Joseph, MI 49085-9659, USA, 1987.
28. SASTRY, S. K. and ZURITZ, C. A. A review of particle behavior in tube flow: applications to aseptic processing. *J. Food Proc. Engr.,* 1987, **10**, 27–52.
29. HODGSON, G. W. and CHARLES, M. E. The pipeline flow of capsules. Part 1. The concept of capsule pipelining. *Can. J. ChE.* 1963, **41**, 155–61.
30. ELLIS, H. S. The pipeline flow of capsules. Part 3. An experimental investigation of the transport by water of single cylindrical and spherical capsules with density equal to that of the water. *Can. J. ChE.,* 1964, **42**, 1–8.
31. SEGRE, G. and SILBERBERG, A. Radial particle displacements in Poiseuille flow of suspensions. *Nature,* 1961, **189**, 209–10.
32. RUBINOW, S. I. and KELLER, J. B. The transverse force on a spinning sphere moving in a viscous liquid. *J. Fluid Mech.,* 1961, **11**, 447–59.
33. SAFFMAN, P. G. The lift on a small sphere in slow shear flow. *J. Fluid Mech.,* 1965, **22**, 385–400.
34. SAFFMAN, P. G. Corrigendum to the paper 'The lift on a small sphere in slow shear flow'. *J. Fluid Mech.,* 1968, **31**, 624.
35. COX, R. G. and BRENNER, H. Effect of finite boundaries on the Stokes resistance of an arbitrary particle. Part 3. Translation and rotation. *J. Fluid Mech.,* 1967, **28**, 391–411.
36. LAWLER, M. T. and LU, P. C. The role of lift in the radial migration of particles

in a pipe flow. In Advances in Solid–Liquid Flow in Pipes and its Application (ed. ZANDI, I.), Pergamon Press, Oxford, New York, 1971, pp. 39–58.

37. DURST, F., MILOJEVIC, D. and SCHONUNG, B. Eulerian and Lagrangian predictions of particulate two-phase flows: a numerical study. *Appl. Math. Modeling*, 1984, **8**, 101–15.

38. SASTRY, S. K. and ZURITZ, C. A. A model for particle suspension flow in a tube. ASAE paper No. 87-6537. American Society of Agricultural Engineers, St. Joseph, MI 49085-9659, USA, 1987.

39. BEAN, P. G., DALLYN, H. and RANJITH, H. M. P. The use of alginate sore beads in the investigation of ultra-high temperature processing. In *Food Microbiology and Technology* (eds Jarvis, B., Christian, J. H. B. and Michener, H. D.), Proc. Intl Cong. Food Micro. and Tech., Parma, Italy, 281–94, Medizina Viva Servizio Congressi S.r.1., 1979.

40. BROWN, K. L., AYRES, C. A., GAZE, J. E. and NEWMAN, M. E. Thermal destruction of bacterial spores immobilized in food/alginate particles. *Food Microbiol.*, 1984, **1**, 187–98.

41. STUMBO, C. R. *Thermobacteriology in food processing*, Academic Press, New York, London, 1973.

42. SASTRY, S. K., LI, S. F., PATEL, P., KONANAYAKAM, M., DOORES, S. and BEELMAN, R. B. A bioindicator for verification of thermal processes for particulate foods. Presented at the 47th Ann. Mtg Inst. Food Technol., Las Vegas, NV, June 16–19, 1987.

43. RAO, M. A. and LONCIN, M. Residence time distribution and its role in continuous pasteurization. Part II. *Lebensm. Wiss. u. Technol.*, 1974, **7**, 14–17.

Chapter 6

PHOSPHATES IN MEAT PRODUCTS

JOHN N. SOFOS

Department of Animal Sciences, and
Department of Food Science and Human Nutrition,
Colorado State University, USA

SUMMARY

The chemical functions of inorganic condensed phosphates (i.e. buffering, sequestering and polyanionic properties) are useful in foods, where they influence water retention, binding, emulsification, coagulation, texture, color and flavor. In meat products, phosphates affect water and fat retention, and meat particle cohesion, texture, color and flavor.

Phosphates are particularly useful in formulations with reduced levels of sodium chloride, where they maintain water and fat binding and product quality. The latest theory has proposed that the mechanism by which phosphates improve water binding is through an increase in protein functionality mainly by alteration of hydrophobic interactions that stabilize the protein structure.

The influence of phosphates on microbial growth appears to be variable and depends on phosphate type and concentration; type and properties of the substrate; pH value; type and level of microbial contamination; heat treatment; other chemical additives; and storage conditions. In general, microbial inhibition varies with phosphate type and chain length, and it may be influenced by phosphate hydrolysis, which will affect chain length. Phosphate hydrolysis is catalyzed by enzymes or heat, which also destroys phosphatases.

The antimicrobial activity of alkaline phosphates increases with increasing chain length, but they are generally less effective than sodium acid pyrophosphate which lowers product pH. Its antimicrobial activity, however, is derived both from reduced pH and from the phosphate ion. Additional studies, however, are needed to examine the interaction of pH and various phosphates on microbial growth.

The antimicrobial activity of phosphates is influenced by heat treatments, and they appear to affect microbial resistance to heat and recovery of heated microorganisms. Existing data, however, are too limited to fully explain these influences. The antimicrobial activity of phosphates is generally increased in presence of other food preservatives, such as sodium chloride, nitrite and sorbate.

Although the mechanism of microbial inhibition is not well defined it is believed that chelation of metal ions by phosphates may play a role in their antimicrobial activity. It is also suggested that phosphates inhibit enzyme activity, and thus can interfere with transport functions, metabolism and activation of toxins.

1. INTRODUCTION

Phosphates are chemical compounds that can act as buffering and chelating agents, and as polyanions influencing the ionic strength of solutions. The chemical properties and reactions of phosphates with food components and other additives influence water and fat retention, food particle cohesion, texture, color, flavor, emulsification, coagulation, curing and possibly preservation of food products.

In general, phosphates are compounds which find diverse uses as functional additives in various foods, including red meat, poultry, seafood, dairy, cereal, bakery, fruit and vegetable products. Depending on food product, type of phosphate and conditions of use, the compounds may improve binding and emulsification of food ingredients; water binding and retention; and enhance or stabilize quality attributes such as color, flavor and texture. The antimicrobial properties of phosphates are not well defined, but the compounds may be useful as components of multifactor preservation systems in certain foods.

In commercial applications it is very difficult to distinguish among the various effects and potential contributions of phosphates and other parameters on product manufacture, quality and preservation. Thus, several effects may be achieved simultaneously, and treatment with phosphates may influence various properties, and contribute several benefits to a single product. Conversely, use of a phosphate may not be contributing all the benefits that it is capable of under optimum conditions.

In meat products phosphates may perform many of their functions, such as water and fat retention, and meat particle cohesion; color development and stabilization; flavor enhancement and inhibition of rancidity; improvement of texture and product handling during manufacture and distribution; and, potentially they may contribute to their preservation. Their usefulness is particularly important in meat products formulated with reduced levels of sodium chloride in order to reduce sodium levels in the human diet. In these products phosphates supplement the functional effects of sodium chloride by improving binding and cooking yields, and potentially shelf-life. Thus, the use of phosphates in meat products in the United States has increased since they were permitted for application in a wider range of items in 1982.[207]

Several review publications have examined the properties and use of phosphates in food products. The most recent of these are those of Hamm,[43] Sofos,[178] Tompkin[201] and Wagner.[215]

2. PROPERTIES OF PHOSPHATES

2.1. General

Phosphates are salts of phosphoric acid with one or more phosphate anions. They are divided into two classes, the simple or orthophosphates and the condensed or polyphosphates. Polyphosphates may be linear, cyclic or cross-linked.[48] The ones used in foods are linear inorganic condensed phosphates (Table 1). The orthophosphates contain a single phosphate anion and are formed by neutralization of phosphoric acid with alkali metal ions, forming monobasic, dibasic, etc. compounds, depending on the number of hydrogen ions replaced with alkali metal ions.[178,215]

Polyphosphates contain two or more phosphate anions and are manufactured by high temperature condensation of orthophosphates to form phosphate chains. The simplest polyphosphate is pyrophosphate, which contains two phosphate anions, while tripolyphosphate contains three phosphate anions (Table 1). Polyphosphates with more than three phosphate anions in their chain are usually called glassy polyphosphates and their average chain length is 6, 12, 22 or more phosphate anions. Most common of these compounds are tetrametaphosphate and hexametaphosphate.

Pyrophosphates and tripolyphosphates are crystalline compounds, while longer chain length condensed phosphates are glassy or amorphous materials. In general, polyphosphates commonly used in foods are linear, unbranched polymers of the general formula $M_{n+2}P_nO_{3n+1}$, where M is

TABLE 1
COMMON PHOSPHATES AND THEIR CHEMICAL FORMULAS

Name	Abbreviation	Formula	Structure
Trisodium phosphate	TSP	Na_3PO_4	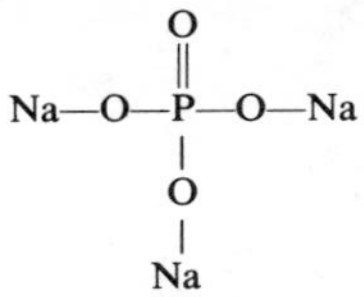 (structure shown)
Tetrasodium pyrophosphate	TSPP	$Na_4P_2O_7$	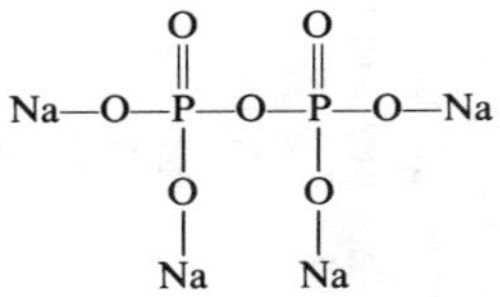 (structure shown)
Sodium acid pyrophosphate	SAPP	$Na_2H_2P_2O_7$	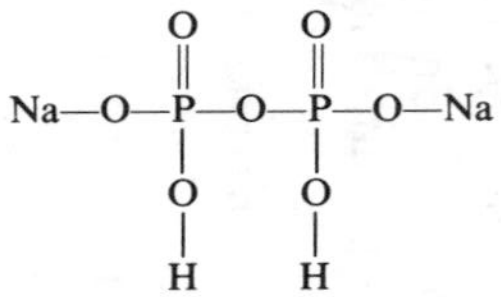 (structure shown)
Sodium tripolyphosphate	STPP	$Na_5P_3O_{10}$	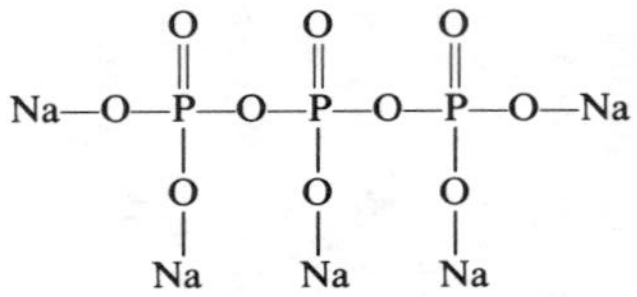 (structure shown)
Sodium hexametaphosphate	SHMP	$(NaPO_3)_{13}$—Na_2O (average)	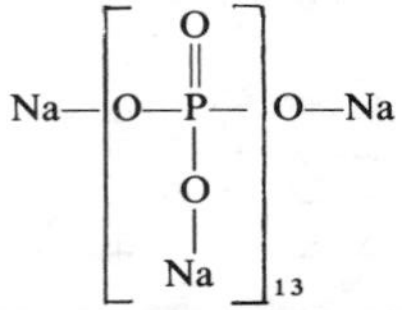 (structure shown)

hydrogen or a monovalent metal ion.[48] Compounds with $n > 5$ are mixtures of polymers with small amounts of cyclic metaphosphates, while polyphosphates with $n \leq 5$ are generally pure. Cyclic condensed phosphates have the general formula $M_nP_nO_{3n}$, and are converted to linear polyphosphates by the action of strong alkali. Cross-linked condensed phosphates or ultraphosphates are found in certain phosphate melts. They have branches in which three oxygen atoms are shared with other phosphate groups. The branching points are readily hydrolyzed in water, which eliminates their presence in biological systems.[48]

Most phosphates are fully neutralized compounds with no hydrogen atoms, and act as alkali in solutions. The most common acidic polyphosphate is sodium acid pyrophosphate which contains two hydrogen atoms per molecule.

Hydrolysis of polyphosphates is the opposite of the production process and can lead to degradation to orthophosphates.[160,161] Hydrolysis is achieved by phosphatase enzymes present in unheated animal products and by heat.[3,45,106,109,112,113,115,123,185]

2.2. Nomenclature

A single phosphate compound may be referred to by several names, which may be as many as 25, as in the case of monosodium phosphate. For example sodium acid pyrophosphate is usually abbreviated as SAPP, but it may also be listed as dibasic sodium pyrophosphate, which is its reference name, or as acid sodium pyrophosphate, disodium dihydrogen pyrophosphate and disodium pyrophosphate. Sodium tripolyphosphate (STPP) is the standard or abbreviated name of pentasodium triphosphate. Tetrasodium pyrophosphate (TSPP) is used for basic sodium pyrophosphate, which is also called sodium diphosphate, sodium pyrophosphate, sodium pyrophosphate tetrabasic and tetrabasic sodium pyrophosphate. Thus, there may be confusion in the literature when different publications use different names for the same compound.

This confusion becomes even greater when commercial names are used for a given compound, and when a commercial name is used for a blend of phosphates. Thus, sodium tripolyphosphate is called Freez-Gard Formula FP-19 by Stauffer Chemical Co., while Freez-Gard Formula FP-88E is a blend of sodium hexametaphosphate, sodium chloride and sodium erythorbate. Sodium hexametaphosphates may be called glassy sodium phosphates, sodium phosphate glass, sodium polyphosphate glassy, Graham's salt, Glass H, Hexaphos or Sodaphos by FMC Corp., metaphosphoric acid sodium salt, SQ Phosphate or Metafos by Monsanto Co., and Vitrafos or Calgon by Stauffer Chemical Co. Sodium tripolyphosphate/sodium hexametaphosphate blends are known as Curafos Formula 11–2, 22–4, Kena Formula FP-28 (Stauffer Chemical Co.), etc. Trade names for sodium acid pyrophosphate include BP Pyro Victor Cream, Taterfos, Perfection or Donut Pyro (Stauffer Chemical Co.), Curacel (Griffith Laboratories), and RD-I (Monsanto Co.). Obviously, products and blends of a given or different companies may be different in properties and composition. This would make them more suitable for specific applications. In addition, many of the names found in the literature may now be obsolete because their production may have ceased due to changes in the industry. Shimp[160] published a table with the various names of phosphate compounds and of commercial blends.

2.3. Chemical

The major chemical functions of phosphates are to control pH by acting as buffers; to sequester metal ions; and, to increase the ionic strength of solutions by acting as polyelectrolytes.[40,159–161,183]

Buffering is the ability of a compound to maintain the pH of solutions at constant levels in the presence of components with different pH values.[215] The good buffering properties of phosphates are useful in stabilizing the pH of foods at levels which are useful in their processing.[40] Orthophosphates are the most potent buffering agents, while pyrophosphate is a good buffer in the pH range (5·5–7·5) of most foods.[178] The longer chain polyphosphates are less effective as buffers in the pH range of foods, and their buffering capacity decreases with increasing chain length.[24,160,183,211]

Chelation is the property of binding metal ions in solutions, and preventing their participation in chemical reactions.[215] This property appears to be related to the antimicrobial, functional and antioxidative properties of phosphates in foods.[178] The orthophosphates do not chelate, the chain polyphosphates are strong chelators, while the cyclic phosphates are weak chelators of metal ions.[210,212] The most stable complexes with metal ions are formed by the longer chain polyphosphates. Shorter chain phosphates, such as pyrophosphate, are better chelators of heavy metal ions (e.g. iron and copper) and their chelating ability decreases with increasing pH values.[59,178,183] Longer chain polyphosphates are better chelators of light metal ions (e.g. calcium and magnesium), and their chelating capacity increases with increasing pH values.[58,160,178,183,215]

Phosphate ions have more than one negative charge and can act as polyanions in a solution. This property allows phosphates to be reactive with food components and molecules, and to affect their surface charge, which makes them important components in food processing.[40] The maximum number of negative charges of orthophosphates is three, depending on pH and concentration. Polyphosphates are stronger polyelectrolytes because they are chains of phosphates. In general, phosphates can interact with polyelectrolytic organic compounds, such as proteins, pectins and starches.[40] By increasing the negative charges on colloidal particles the phosphates can act to deflocculate, disperse or emulsify relatively insoluble food components in aqueous systems.[40] Phosphates increase the ionic strength of solutions, and can bridge two or more positively charged sites, and thus, bind components or particles, which can lead to precipitation.[183] By attaching one end of their chain to a positively charged site and the others to water molecules, phosphates can also maintain particles in solution.[215]

2.4. Functional

The functional properties of phosphates in food systems are due to their basic chemical properties, namely, the pH buffering ability, their chelating properties and their polyanionic behavior. Beneficial contributions in processed foods include water and fat retention, and meat particle cohesion; development and stabilization of color; prevention of oxidative rancidity; improvement of texture; prevention of coagulation; emulsification; leavening; enhanced curing; easier processing; and, nutritional enhancement.[21,160,178,201]

The beneficial effects of phosphates in food systems are derived from their interaction with natural constituents and additives in the food. Thus, the functionality of a given phosphate depends on and will vary with, the properties of individual food systems. Use of phosphates in various foods must be decided on the basis of the benefits desired and the specific properties of each food. Phosphates may contribute opposite effects, depending on their properties, type of food, and processing conditions.

Examples of the functions and uses of phosphates in food systems demonstrate the diversity of their activity, which, as indicated, depends on the properties of individual foods, phosphate types, conditions of processing and other chemical additives.[161] A well-known use of phosphates is to increase moisture retention and improve cooking yields and quality of processed muscle foods. Retention of soluble nutrients, such as proteins, vitamins and minerals may also improve the nutritional quality of the products.

Phosphates themselves are not able to emulsify mixtures of oil and water, but in processed cheese products, which contain protein, phosphates such as disodium phosphate act as emulsifiers. In these products, the proteins are the actual emulsifiers, but the phosphate interacts with the proteins and increases their emulsifying capacity.

Filtration of whey and concentration of lactose is facilitated by phosphates, because they precipitate whey proteins. In this application, the whey is acidified to the isoelectric point of its proteins, and a glassy polyphosphate aggregates the protein molecules and speeds up coagulation, precipitation and separation.

In contrast to precipitation, glassy polyphosphates can also stabilize proteins and prevent their coagulation. This property is useful in the manufacture of pasteurized liquid egg products. Addition of phosphates to the eggs, before heat treatment, not only prevents coagulation of egg whites, but it also enhances microbial destruction at lower temperatures.

Acidic phosphates, such as sodium acid pyrophosphate, are useful in leavening of bakery products, although they have no direct leavening activity. Their usefulness is based on their ability to neutralize sodium bicarbonate, which releases carbon dioxide and results in leavening.

Alkaline and acidic polyphosphates, such as sodium tripolyphosphate and sodium acid pyrophosphate, respectively, chelate pro-oxidant metal ions (e.g. iron and copper) and prevent oxidative deterioration of foods. Thus, phosphates act as antioxidants, even though they are unreactive toward oxygen.

Other uses of phosphates are to stiffen whipped toppings and prevent release of water; to stabilize proteins in imitation dairy products and dry milk powder; to prevent churning in ice cream; to prevent gelling in evaporated milk; to control tartness in dry beverages; to buffer gelatin desserts; to improve properties of starch in breakfast cereals; to prevent discoloration in fruits and vegetables; and, to contribute nutrients in the production of baker's yeasts and other microorganisms.

In general, since phosphates contribute several benefits to various food products, selecting a phosphate only on the basis of a single desired effect is not always the best approach. For example, an orthophosphate may contribute buffering capacity to a food system, but a pyrophosphate may contribute both buffering and additional benefits, such as inhibition of oxidation and microbial growth.[160,161]

2.5. Health

Phosphorus and phosphates are essential in life and therefore phosphates used in food processing should be considered as safe additives, when used at proper concentrations. Actually the United States Food and Drug Administration considers phosphates used in processed foods to be generally recognized as safe (GRAS) food ingredients. Potential health problems from consumption of excessive phosphate concentrations include short-term abdominal distress and bone calcium mobilization.[24,67,142] Such problems are avoided, however, by minimizing use of phosphates in the diet.[203]

As indicated earlier, phosphates may improve the nutritional quality of treated foods by preventing losses of nutrients during processing. One concern, however, is their effect on the bioavailability of minerals in the diet. Increased consumption of dietary phosphate (i.e. orthophosphate) has been reported to decrease the absorption of iron and zinc in animals and humans,[171,234,237] but had little or no effect on zinc and calcium utilization by humans.[171,237] Hexametaphosphate has decreased absorption of calcium[236,237] and iron,[234,235,237] are tripolyphosphate has decreased absorption of iron.[235] Some studies, however, indicated that both hexametaphosphate and tripolyphosphate increased absorption and utilization of zinc.[234,235] Studies involving *in-vitro* peptic and peptic–pancreatic digestions indicated that orthophosphate depressed zinc availability, while the polyphosphates enhanced it.[233] Additional effects of phosphates on calcium and iron solubility during simulated digestion are shown in Table 2.

TABLE 2

EFFECT OF PHOSPHATES (1%) ADDED TO GROUND BEEF ON SOLUBLE CALCIUM AND IRON, AND IONIC IRON FOLLOWING SIMULATED GASTRIC AND GASTROINTESTINAL DIGESTIONS

	Soluble (%)		Ionic (%)
Conditions	Calcium	Iron	Iron
Gastric digestion			
Control	100	100	100
Orthophosphate	41	155	138
Tripolyphosphate	44	127	undetectable
Hexametaphosphate	39	205	14
Gastrointestinal digestion			
Control	100	100	100
Orthophosphate	146	92	86
Tripolyphosphate	428	92	86
Hexametaphosphate	400	100	51

Data extracted from Zemel,[233] *Journal of Food Science*, 1984, **49**, 1562. Copyright © by Institute of Food Technologists.

Mineral bioavailability *in vivo*, however, will depend on type and level of phosphate, type of food, and other variables.

2.6. Methods of Application

Methods for adding phosphates to meat, poultry and seafood products include immersion, spraying, tumbling or injection with phosphate solutions, and direct addition of dry salts.[161,183] The method of choice in specific applications is based on the type and size of the product and on procedures involved in processing.

Immersion or spraying procedures are simple, but they cannot be used in large meat items because penetration of the solution is limited. In contrast, these products can be treated by injection. Tumbling of meat pieces in a solution of phosphate and other ingredients is desirable when the meat pieces should be coated with extracted proteins to enhance binding. Some of the phosphate solution may be injected into the meat before tumbling. Addition of phosphates in the dry form can be applied in chopped or comminuted meat products, where it is feasible.

3. NONMICROBIAL EFFECTS

3.1. General

Polyphosphates have been used in the United States since 1952 for their beneficial effects in cured pork products, such as ham and bacon. In these products phosphates improve retention of moisture and meat cohesion; increase cooking yields; improve quality; and facilitate product slicing without crumbling.

Use of phosphates in meat products in the United States increased in the 1980s due to liberalization of the regulations, which permitted their use in a wider variety of meat products.[207] The change in the regulation and the subsequent increased use of phosphates in meat products in the United States was due to attempts to reduce sodium chloride levels, and thus, lower sodium consumption and potential development of human hypertension.

Specific beneficial effects of phosphates in meat products are enhanced water retention, meat particle cohesion and fat binding; retardation of oxidative rancidity; and, development and preservation of color.[161,178,183] They also reduce viscosity ('slacken') or soften meat batters,[46,74,173] which prevents overheating during mechanical treatment, and improves automated handling and extrusion.[202]

In general, use of phosphates in red meat, poultry and seafood products reduces cooking losses and purge; increases juiciness, improves color and flavor stability; improves texture and tenderness; increases protein extraction; improves sliceability; controls struvite in canned seafood; etc.[97,178] Potential drawbacks on meat product texture (e.g. tough, rubbery) and flavor (e.g. metallic, astringent, soapy) can be avoided through optimization of levels and conditions of use.[178,203]

Numerous studies have established the usefulness of phosphates in meat products through their influence in improving water retention and meat particle cohesion,[41,42,50,72-74,102,162,178,186,187,203-205] color,[195] and flavor.[2,69,145,169] Although use of a given phosphate in muscle foods may contribute many beneficial effects, it is difficult to isolate and quantitate specific functions and contributions in complex meat systems. This becomes even more complex because the activity of phosphates in meat products is affected by several factors including type, composition and pH of product; phosphate type and concentration; type of processing; and, presence of other additives (Tables 3 and 4).

3.2. Binding

The most important function of phosphates in meat products is to enhance water retention and meat particle cohesion, which improves cooking yields

TABLE 3

INTERACTION OF pH, SODIUM CHLORIDE AND SODIUM ACID PYROPHOSPHATE (SAPP) IN FRANKFURTER BATTERS

Property	SAPP (%)	1·5% NaCl		2·5% NaCl	
		pH 5·5	pH 5·8	pH 5·5	pH 5·8
Water exudate (%)	0	22·6	6·7	5·7	2·3
	0·25	3·3	1·3	0·3	0·0
Fat exudate (%)	0	3·7	0·3	0	0
	0·25	0	0	0	0
Gel strength (kg)	0	0·39	0·61	0·67	0·64
	0·25	1·00	0·77	0·85	0·87

Data extracted from Whiting,[224] *Journal of Food Science*, 1984, **49**, 1355. Copyright © by Institute of Food Technologists.

TABLE 4

INTERACTION OF pH, SODIUM CHLORIDE AND PHOSPHATES (PYROPHOSPHATE, PP; HEXAMETAPHOSPHATE, HMP) IN RESTRUCTURED BEEF ROLLS

Product pH	NaCl (%)	Phosphate	Percentage increase due to phosphate	
			Cook yield	Tensile strength
5·83	0·8	PP	16·5	32·3
	2·2		43·0	57·1
	0·8	HMP	7·6	16·6
	2·2		23·0	37·2
6·17	0·8	PP	19·7	44·6
	2·2		48·0	78·6
	0·8	HMP	8·2	21·9
	2·2		31·5	41·7

Phosphates added at ionic strength of 0·055.
Data extracted from Trout and Schmidt,[205] Journal of Food Science, 1986, 51, 1416. Copyright © by Institute of Food Technologists.

and product quality.[24,97,159-161,178,183] The influence of phosphates in improving water and meat binding is enhanced when used in combination with sodium chloride.[7,41,42,50,64,65,82,86-88,94-97,130,158,162,164,204,205,214] Phosphates also improve functionality when lower quality or aged meats are used in sausage manufacture.[17]

Since meat consists of high amounts of water (65–75%), which can be released during processing treatments and storage, phosphates are useful because they reduce or eliminate these losses.[159-161] Prevention of water losses results in economic and quality benefits, because processing yields are higher, water soluble nutrients remain in the product, and the moistness, textural and flavor qualities of the product are maintained.

Binding of meat pieces is important in products manufactured from chunks, flakes, chopped or finely comminuted meat. Binding of meat pieces yields cohesive products of acceptable yield and quality, and of the appropriate identity. The function of phosphates in these products is on the salt-soluble muscle proteins, which form a viscous coating on the surface of meat particles. Subsequent heating of the products coagulates the proteins, which form a matrix that binds the meat pieces and entraps other ingredients, forming cohesive and uniform products.[1,150,161] These effects of phosphates and muscle proteins occur in the meat products from various species, including buffalo and goat meats.[78]

3.2.1. Mechanisms

It has been reported that phosphates improve water and meat binding by increasing pH and ionic strength; dissociating actomyosin to actin and myosin; chelating metal ions; and, by binding to meat proteins.[41,42,151,203-205] The relative importance of each of these effects, however, is still under discussion.

Increases in ionic strength and pH are probably the most important factors in improving meat particle-to-particle cohesion and cooking yields.[42,50,138,204,205] Trout and Schmidt[204,205] reported that most of the variation in binding and cooking yields of restructured beef rolls could be explained by changes in pH and ionic strength. Ionic strength explained 53–59% of the variation in cooking yield, while pH accounted for 25–30% of the variation. A variation of 5–9% was due to the phosphate molecule.[205] A pH value of 6·0 and a total ionic strength of 0·60 resulted in maximum binding and cooking yield of beef rolls.

Increases in pH (Tables 3 and 4) have been considered important in increasing water retention in meat since the early years of phosphate usage.[158,162,186,187] As the pH of the meat is raised above the isoelectric point of muscle proteins (5·4), the proteins unfold and react with polar water molecules to compensate for their increasingly unbalanced internal charge.[161] Thus, raising the pH of meat with alkaline phosphates is beneficial in improving functionality in meat products.

The importance of pH changes (Table 5) by phosphates in meat products, which are moderate (i.e. 0·1–0·7 in uncooked and 0·05–0·3 in cooked products) and depend on phosphate type and concentration, has been disputed.[42,50,144,204,205] Other compounds that raise the pH, such as sodium hydroxide, do not increase water binding to the same extent as alkaline phosphates.[74] Orthophosphates, which have lower chelating power and insignificant polyanionic activity compared to polyphosphates, also increase the pH, but their influence on water and meat binding is negligible.[159,205] Buffered tripolyphosphate and pyrophosphate mixtures can improve water retention even without changing the pH.[161] The extent of pH increase by alkaline phosphates is variable and depends on buffering capacity of meat, type of product, cooking, type and level of phosphate, and concentration of sodium chloride.[78,86,92,117,140,176–178,224] Increases in pH of comminuted beef with tripolyphosphate were lower with increased levels of sodium chloride,[17] and with meat of higher initial pH values. Sodium hexametaphosphate causes smaller increases in pH compared to tripolyphosphate.[108] Sodium acid pyrophosphate results in pH reduction by 0·1–0·3 units (Table 6), but it also improves binding at both higher and lower pH values.[92,93,224] Thus, raising the pH of meat with alkaline phosphates is beneficial in improving functionality, but it is not the only mode of action by phosphates. The importance of pH increases in meat

TABLE 5
EFFECT OF PHOSPHATES (0·5%) ON pH OF MEAT BATTERS

| Treatment | Frankfurters | | Bratwurst |
	Raw	Cooked	Cooked
Control	6·01	6·23	6·45
Pyrophosphate	6·91	6·81	6·80
Tripolyphosphate	6·85	6·71	6·65
Hexametaphosphate	6·55	6·57	6·38
Acid pyrophosphate	6·12	6·23	6·10

Data extracted from Knipe et al.[74] and Molins et al.,[115] Journal of Food Science, 1985, **50**, 1017 and 1985, **50**, 876. Copyright © by Institute of Food Technologists.

TABLE 6

INFLUENCE OF SODIUM ACID PYROPHOSPHATE (SAPP) ON FRANKFURTER pH

SAPP (%)	Raw product pH	Cooked product pH
0·00	6·32	6·22
0·09	6·28	6·20
0·19	6·16	6·02
0·28	6·14	6·02
0·37	6·12	5·98
0·46	6·05	5·95

Data extracted from Hargett et al.,[46] Journal of Food Science, 1980, **45**, 905. Copyright © by Institute of Food Technologists.

binding may vary with meat products, pH range and type of phosphate.

Increasing the ionic strength of meat increases water retention. All phosphates increase ionic strength, but their effect on meat cohesion is variable (Table 4).[42,44,204,205] The effect of individual phosphates on ionic strength is also variable.[205] Increases in ionic strength produced by phosphates are difficult to determine due to variations in their dissociation. Increasing the chain length of phosphates causes a decrease in dissociation, which results in lower ionic strength values with longer chain polyphosphates compared with similar amounts of shorter chain polyphosphates.[205]

Other salts also contribute to the ionic strength of meat products. At ionic strength values above 0·15 (0·8% sodium chloride) polyphosphates increase the functional properties of meat more than other salts causing comparable changes in pH and ionic strength.[36,139,161,205] The extent of this synergistic effect, however, decreases linearly as the chain length of the polyphosphate increases.[205] Reduced phosphate dissociation (and ionic strength) at longer phosphate chain lengths is at least a partial reason for this reduction in functionality.

Speculations to explain the synergistic interaction of sodium chloride and phosphates at certain ionic strengths have included potential dissociation of actomyosin to actin and myosin by phosphates,[7,50,159-161,229,230] and binding of phosphates to meat proteins.[42,205] Both of these explanations have been disputed as being of importance in meat binding.[203-205]

Meat proteins are extracted by mechanical action and ionic strength during processing. Dissociation by phosphates and solubilization by ionic strength is believed to increase their binding properties. Heat treatment then coagulates the protein matrix to form a gel that binds the particles and seals the pores preventing losses of water, fat and nutrients.[158,159] Several studies have reported increased protein extractability in meat systems with sodium chloride and phosphates.[135-137,165-168,196,197] Actually, inclusion of phosphates allows reduction of sodium chloride without reducing protein extractability,[33,135] and increases muscle fiber swelling in the raw state.[127]

Shorter chain length polyphosphates (e.g. pyrophosphate, tripoly-phosphate) were thought to react with actomyosin and show increased activity in the presence of sodium chloride. Longer chain hexameta-phosphates were thought to bind directly to myosin, but binding was inhibited at high levels of sodium chloride.[229] However, the importance of protein–phosphate binding and actomyosin dissociation by phosphates in meat functionality has been questioned.[205] The polyanionic nature of phosphates permits them to attach to protein sites and attract more water molecules,but there is no direct relationship between the extent to which phosphates bind to meat proteins and their effect on functionality.[203] Meat enzymes also can hydrolyze polyphosphates to orthophosphates, which can reduce their potential water binding activity through reaction with meat proteins.

Dissociated actin and myosin exhibit strong interaction with water to compensate for the lost interaction between themselves.[161] Actomyosin, however, is dissociated only by pyrophosphate, while other polyphos-phates will dissociate actomyosin only when hydrolyzed to pyrophosphate. The rate and extent of polyphosphate hydrolysis may be variable with types of products and conditions of processing.[109,112,113,115] Although direct evidence is lacking, only tripolyphosphate may hydrolyze to pyrophos-phate at a sufficient rate before product cooking to cause dissociation of actomyosin.[205] Polyphosphates, not believed to undergo rapid hydrolysis, have increased functionality in beef rolls[205] and other products. Thus, protein–phosphate reaction and actomyosin dissociation by phosphates may be involved in functionality of only certain products formulated with specific phosphates.

Another potential mode of action of phosphates in increasing meat functionality is through their chelating activity on cations, such as calcium, magnesium and zinc.[41,42,227] The speculation is that the cations are chelated and thus removed from the proteins, which opens up their structure and improves their water binding properties. This theory has also been disputed by several researchers.[50,57,79,205] Polyphosphates did not reduce the amount of calcium and magnesium ions bound to protein. Sixty percent of the calcium and 20% of the magnesium naturally present in meat were firmly bound and not available to react with added phosphates.[57] In addition, meat treated with a chelating agent, such as ethylene diamine-tetracetic acid, showed no increase in water retention,[50,158] and addition of calcium and magnesium to meat did not reduce water binding,[7,24] Also the variation among phosphates in their effectiveness on meat binding suggests that metal ion chelation may not be a major factor in meat binding.[159]

Based on their results on beef rolls with different phosphates, ionic strengths and pH values, Trout and Schmidt[205] have advanced a theory to explain the activity of phosphates as meat binders. Based on regression equations it was found that 93% of the variation in cooking yields and 86% of the variation in tensile strength of beef rolls could be explained in terms of pH, molar sodium chloride and molar phosphate concentrations. This confirmed their conclusion that phosphates increase protein functionality mainly by altering hydrophobic interactions that stabilize the protein structure. This was supported by the linear decrease in the effectiveness of phosphates with increasing chain length, which must have been due, at least in part, to changes in hydrophobic interactions. If increased protein functionality in presence of phosphates was due to changes in electrostatic interactions, all treatments of the same ionic strength and pH (even without phosphate) would have increased functionality to the same extent. Regression analysis of the results showed that the decrease in functionality with increasing phosphate chain length at equal weight concentrations of all phosphates was linearly related to the decrease in molar phosphate concentration. It should be noted that with increasing chain length of phosphate, molar concentrations decrease at equal weight concentrations. Thus, the linear decrease in effectiveness may have been due to either or both of these changes, that is increased chain length and decreased molar concentration.

Increased protein functionality by phosphates through changes in hydrophobic protein interactions was also supported by the lack of an effect by pH on the effectiveness of phosphates, which eliminates the potential for phosphates to be active through alteration of electrostatic interactions. Thus, the hydrophobic effects of the phosphates had the predominant effect on functionality.[205]

Trout and Schmidt[205] also indicated that the synergistic increase in functionality produced by phosphates and sodium chloride at ionic strengths above 0·15 can be explained by the opposing effects on electrostatic and hydrophobic interactions. At low ionic strength, phosphates change the electrostatic interactions, which alters protein conformation. As the ionic strength increases the electrostatic effects of phosphates decrease, because of increased ion shielding and decreased charge on the phosphates caused by higher sodium chloride levels associated with higher ionic strength. As the electrostatic effects of phosphates are reduced, the hydrophobic effects become dominant, and potentially increase the functional properties of meat proteins.

3.3. Quality

Although the main objective of adding phosphates to meat products is to improve water and meat binding, the compounds also improve the chemical, textural and sensory properties of the products. The antioxidant properties of phosphates are especially important because they influence both flavor and color stability in meats.[159-161,178,215]

Oxidative reactions are major causes of color and flavor deterioration during storage of muscle foods. Rancid off-flavors in meat, poultry and seafood are due to lipid oxidation, which is catalyzed by metal ions, such as iron and copper.[24,161] Oxidation of myoglobin also results in undesirable meat colors. Phosphates, through their chelating properties, prevent these metal ions from participating in oxidation reactions, and, thus inhibit development of rancidity and stabilize meat color.[159]

Several studies have reported on the inhibition of oxidative rancidity (Table 7) and improvement of flavor and color with phosphates in meat products.[14,16,49,54,68,99-101,143,153,169,170,200,223] Phosphates have improved color and flavor in frankfurters,[92,93,178] restructured beef and pork

TABLE 7

EFFECT OF SODIUM CHLORIDE AND SODIUM TRIPOLYPHOSPHATE ON THE RANCIDITY (TBA VALUE, mg MALONALDEHYDE/kg) OF PORK SAUSAGE PATTIES

Sodium chloride (%)	Sodium tripolyphosphate (%)	TBA value
0	0	0·64
0	0·375	0·25
0·5	0	0·78
0·5	0·375	0·45
1·0	0	1·78
1·0	0·375	0·61
1·5	0	1·35
1·5	0·375	0·54

Data extracted from Matlock et al.,[100] Journal of Food Science, 1984, **49**, 1363. Copyright © by Institute of Food Technologists.

nuggets,[53] frozen restructured beef steaks,[83,108] frozen beef patties,[116] pork sausage patties,[68,100,101] and freeze-dried beef.[119]

In addition to inhibition of rancidity, flavor improvement with phosphates has been attributed to retention of proteins during cooking.[24] Phosphates also counteract the detrimental effect of sodium chloride on the color of uncured meat products.[12,54,83,108,116,153] In cured meats, nitrite also performs antioxidant functions. The inhibition of discoloration by phosphates has been attributed to prevention of lipid oxidation, since its by-products increase the rate of discoloration.[16,38,55] Increases in pH and chelation of metal ions by alkaline phosphates have also been suggested as reasons for inhibition of lipid oxidation and discoloration.[87-89]

The influence of phosphates on flavor and color of meat products depends on the type of product and the type of phosphate. Sodium tripolyphosphate inhibited rancidity in frozen restructured beef steaks, while sodium hexametaphosphate and sodium acid pyrophosphate did not reduce thiobarbituric acid values.[108] Tripolyphosphate reduced the rate of warmed-over-flavor development in reheated pork roasts, but was ineffective in reheated beef roasts.[169] Hexametaphosphate and tripolyphosphate improved color scores of restructured beef steaks, while sodium acid pyrophosphate was less effective.[108] In addition, sodium acid pyrophosphate did not increase development of cured meat color in frankfurter-type products.[46,195]

Phosphates can reduce or enhance the rate of the curing process, and, thus affect development of cured meat color, depending on changes in product pH. If automated processing is more rapid than the reaction between nitrite and myoglobin, color development may be incomplete.[161] Alkaline phosphates have inhibited color development in rapidly processed frankfurters by increasing their pH. In normally processed frankfurters, however, alkaline phosphate resulted in development of more stable color.[187] Rapid processing of frankfurters can be facilitated with sodium acid pyrophosphate, which lowers the pH to increase the rate of the curing process. The pyrophosphate ion, however, counteracts the detrimental effect of the lower pH on product binding.[10,149]

Use of phosphates in meat products can also improve texture and other sensory attributes.[46,69,82,92,93,108,139,141,178,195,204,205] Specific quality attributes improved by phosphates include firmness, fracturability and hardness by tripolyphosphate in frankfurters;[69] tenderness, juiciness and flavor intensity in pork and beef roasts injected with tripolyphosphate;[169] tenderness in beef and pork;[12,52,66,107] and slice cohesiveness and consistency in pork liver loaf.[131]

Phosphates, such as hexametaphosphate and pyrophosphate can restore tenderness in hot processed pork loins.[51] Injection of the loins with phosphate solutions within one hour *post mortem*, also increased juiciness, but it increased thaw drip, reduced browning during cooking, resulted in less desirable flavor, and increased microbial contamination.

Sodium acid pyrophosphate has improved texture of frankfurters by increasing springiness, hardness and awareness of skin, and by decreasing oiliness.[46] The same acidic phosphate has also improved flavor by enhancing beef, salt, smoke and seasoning flavors, and by diminishing fatty flavors and mouthfeel.

3.4. Poultry Products

Phosphates are used in many types of further processed poultry products, including loaves, ham-type products, frankfurters, pastrami and self-basting turkeys. As in red meats, sodium tripolyphosphate is the most commonly used compound in poultry products, while sodium acid pyrophosphate can be included in the formulation to moderate changes in product pH.

Phosphates in poultry products improve binding, cooking yields, quality and, possibly, shelf-life.[161] Product quality is enhanced through improvement of juiciness, firmness, sliceability, and prevention of oxidation. Injection of rapidly cooled and cut-up poultry carcasses with phosphate solutions reduces drip losses during storage, and prevents toughening.[65,132]

Phosphates increase protein extraction and viscosity of turkey meat batters,[136] and moisture retention, cohesiveness, springiness and chewiness, and flavor quality of other poultry products.[9,32,90,91,231]

3.5. Seafood Products

Phosphates are capable of providing to seafood products benefits similar to those introduced in red meat and poultry items. Actually seafoods constitute a potential growth area for phosphate applications.[160]

Advantages of using phosphates in seafood include increased cooking yields, reduced thaw and refrigerated storage weight losses, improved texture, and resistance to oxidative changes.[228] Use of phosphates (e.g. sodium tripolyphosphate) and sodium chloride in breaded fish patties improves texture by increasing firmness, moistness and flavor scores.[2,11,19]

Phosphates are used by the shrimp industry to facilitate processing and improve quality, since they provide a lubricating action, which enhances mechanical removal of shells. A potential problem of treating fresh seafood with phosphates is extraction of protein and formation of an unaesthetic slimy surface coating. The problem can be avoided by breading and cooking, where it also improves binding of the batter on the surface of the fish. The problem can also be controlled by modifying the treatment conditions.[160]

Production of fish paste (i.e. kamaboko) to be used in surimi products depends on salts (e.g. sodium chloride, tripolyphosphate) to extract the fish proteins, which provide water holding and binding.[71] Phosphates also act synergistically with carbohydrates to stabilize the functional properties of frozen or freeze-dried surimi.[129]

Phosphates also prevent formation of struvite crystals in canned fish, such as tuna. Struvite crystals are harmless, but they are undesirable, because they look and feel like glass. Sequestration of magnesium by a phosphate (e.g. sodium acid pyrophosphate) prevents their formation.[160,161]

3.6. Low Sodium Meat Products

The possible involvement of sodium consumption in the development of human hypertension has prompted public health and regulatory authorities to recommend reduction in the dietary intake of sodium chloride. Since certain meat products contain relatively high amounts of sodium, they are major targets for lowering ingoing levels of sodium chloride.[175,178,179] As sodium chloride levels are reduced, however, meat cohesion, water retention, product quality and shelf-life may be compromised.[173,174,176,177,180] The meat industry has, therefore, turned to the use of various phosphates, which can partially replace sodium chloride and restore product quality and identity.[68,92,93,140,141,154,176,177,204,205,224]

Exact levels of sodium chloride required for acceptable product binding and quality depend on factors, such as raw material and its pH, fat levels and type of product.[173,174,180,205] Sodium chloride levels of less than 1·5%, however, result in unstable meat batters.[175,178,179] Various phosphates can improve the properties of these products. It is also important to know whether phosphates will also provide antimicrobial activity in these low sodium chloride meat products.[178] Use of potassium instead of sodium salts of phosphates can result in even lower levels of sodium in these products.

3.7. Phosphate Types and Levels

Food grade phosphates permitted for use in processed meats in the United States include monosodium and monopotassium phosphate; disodium and dipotassium phosphate; tetrasodium and tetrapotassium pyrophosphate;

sodium and potassium tripolyphosphate; sodium metaphosphate; sodium hexametaphosphate; sodium acid pyrophosphate; and their blends.[207] The United States meat industry uses large amounts of phosphates annually.[13] The most commonly used phosphate and the major component in commercial blends is sodium tripolyphosphate. The compound is used widely because, in addition to being effective in meat binding, it is more soluble in pickle solutions and is less likely to form calcium and magnesium precipitates in hard water. Other phosphates commonly used in blends are sodium hexametaphosphate, tetrasodium pyrophosphate and sodium acid pyrophosphate.

It is very important to select the appropriate phosphate types and blends for specific applications, since their effectiveness on functionality and antimicrobial activity is variable. Several studies have indicated that the most influential phosphate in improving functionality in meat products is pyrophosphate, followed in decreasing order of effectiveness by tripolyphosphate, tetrapolyphosphate, hexametaphosphate and orthophosphate.[7,50,72,162,163,204,205] There are exceptions to this generalization indicating tripolyphosphate being more effective than the others.[44] A blend of pyrophosphate and tripolyphosphate was found at least as effective in reducing processing shrinkage in hams as tripolyphosphate.[214] Orthophosphates are ineffective in water binding,[159,183,205] while sodium acid pyrophosphate, which is the most effective as an antimicrobial agent, also improves functionality.[178,215] Several studies have demonstrated improved cooking yields, binding and texture in meat formulations with the acidic sodium acid pyrophosphate (Table 3), which is a curing accelerator.[92,93,224] Others, however, have reported increased weight losses during cooking in products formulated with sodium acid pyrophosphate.[46,195] Thus, the actual effect of sodium acid pyrophosphate on product functionality may depend on type of product and its pH. The antimicrobial properties of the compound,[5,114,178,215] however, indicate that it should be considered as an ingredient of phosphate blends, especially in reduced sodium chloride meat formulations that may need additional antimicrobial activity.

The United States regulations permit presence of 0·5% total phosphate in finished meat products. Thus, meat processors usually add levels of about 0·4% in the formulation. Several studies, however, have indicated that for maximum functionality in meat formulations containing 1–2% sodium chloride, a phosphate level in the range of 0·1–0·3% should be adequate.[53,108,139,205,231] Phosphate levels providing maximum antimicrobial activity are not well defined.

Both the antimicrobial and functional effects of phosphates depend on

TABLE 8
DEGREE OF DISSOCIATION OF PHOSPHATES

Phosphate	Chain length (n)	Concentration (%, w/v)	Molar concentration $\times 10^3$	Dissociation (%)
Disodium phosphate	1·0	0·15	10·5	91
		0·45	31·6	92
Tetrasodium pyrophosphate	2·0	0·15	5·6	89
		0·45	16·9	88
Sodium tripolyphosphate	3·0	0·15	4·0	83
		0·45	12·1	81
Sodium tetrapolyphosphate	5·7	0·15	2·3	64
		0·45	9·3	63
Sodium hexametaphosphate	12·8	0·15	1·1	45
		0·45	3·3	47
Sodium phosphate, glassy	20·8	0·15	0·7	38
		0·45	2·1	38

Data extracted from Trout and Schmidt.[206] With permission. Copyright ©, 1986, American Chemical Society.

the concentration of sodium chloride present in the formulation. Product binding, texture and flavor are improved by lower phosphate levels when used in the presence of sodium chloride.[153,231] Doubling the level of phosphate was more effective in increasing meat batter pH and soluble protein level, but its effect on batter stability was lower than that caused by doubling of the sodium chloride concentration.[72,73] Actually, increasing the sodium chloride concentration from 0·75% to 1·5% reduced the relative effect of phosphates on batter stability.

In general, phosphates do not improve meat functionality when the level of sodium chloride is below 0·8%,[7,50] higher than 2·0%[187] or in the range of 1·25–1·50%, if the pH is higher than 6·0–6·3.[138,204,205] The lack of improvement in functionality at intermediate and higher sodium chloride levels and at high pH is due to the increased ionic strength which results in maximum functionality.[173,205]

The dissociation of phosphates is also influential on their usefulness in meat systems. The degree of dissociation of phosphates decreases as the chain length increases (Table 8), and the rate of this decrease is proportional to the square of the chain length of the phosphate.[206] The values for the degree of dissociation are not affected by phosphate concentration, and can range from 91% for the shortest chain length (1·0) to 38% for the longest (20·8).[206] As indicated earlier, the degree of dissociation determines the ionic strength, which may explain the lower activity, as functional ingredients, of longer chain polyphosphates.[92,205]

4. ANTIMICROBIAL EFFECTS

4.1. General

Published research on the antimicrobial activity of phosphates in culture media and in food systems is relatively limited, and has been reviewed by Sofos,[178] Tompkin[201] and Wagner.[215] Furthermore, inclusion of phosphates in meat formulations and the levels used are based on their usefulness as functional rather than antimicrobial additives. Nevertheless, their potential influence on meat preservation should be considered, especially in meat products with reduced levels of sodium chloride, which may be more susceptible to microbial spoilage and toxicity problems.

Existing studies on the antimicrobial activity of phosphates often present conflicting conclusions. Several studies have indicated that certain phosphates may delay microbial growth in meat, poultry, seafood and in culture media.[8,25,30,31,39,62,95,96,105,147,182,192] Others, however, have reported that phosphates did not delay, or even stimulated, microbial growth.[20,61,104,199]

In general, the antimicrobial behavior of phosphates should depend on phosphate type and concentration; pH value; type of substrate; heat treatment; type and level of microbial contamination; other additives; and, storage conditions (Table 9).[178]

4.2. Culture Media

Studies employing culture media as substrates are useful in indicating antimicrobial properties, mechanisms of action and interactions of ingredients used in food processing.[201] Several studies with culture media have demonstrated inhibition of various microorganisms by phosphates.

Specific spoilage and pathogenic microorganisms inhibited by phosphates in culture media include pseudomonads,[25] lactic acid-producing bacteria,[110,188] yeasts and molds,[192] *Staphylococcus aureus*,[63,70,110,182,198] *Salmonella typhimurium*,[110,182] *Moraxella-Acinetobacter*,[30] *Streptococcus faecalis*[70,182] *Bacillus subtilis*,[37,70] *Clostridium bifermentans*,[70] *Clostridium sporogenes*[182] and *Clostridium botulinum*.[152,155,215]

As indicated above conclusions with culture media and foods on the relative effectiveness of various phosphates as antimicrobial agents are often conflicting. According to one study,[110] tetrasodium pyrophosphate was highly inhibitory to microbial cultures in culture media, followed by sodium tripolyphosphate and sodium hexametaphosphate, while sodium acid pyrophosphate was ineffective, or even enhanced microbial recovery. Other studies, however, have found sodium acid pyrophosphate more effective as an antimicrobial agent in foods and culture media than tetrasodium pyrophosphate, tripolyphosphate and other phosphates.[178,215]

TABLE 9

RECOVERY OF *MORAXELLA-ACINETOBACTER* CELLS ON PLATE COUNT AGAR AFTER 7 DAYS AT 32°C IN PRESENCE OF SODIUM CHLORIDE AND ORTHOPHOSPHATE

NaCl (%)	Orthophosphate (%)	Recovery (%)
0	0·5	98
0	1·0	65
0·4	0·25	99
0·8	0·20	14
0·8	0·50	0

Data extracted from Firstenberg–Eden *et al.*,[30] *Journal of Food Science*, 1981, **46**, 579. Copyright © by Institute of Food Technologists.

Although microbial proliferation may not be inhibited by phosphates, under certain conditions, they may extend product shelf-life by suppressing metabolic activity of growing cultures. Examples of such phosphate activity include reductions in the rate of histamine production by *Proteus morganii* and *Klebsiella pneumoniae* (Table 10);[193] inhibition of protease secretion by *Aeromonas hydrophila*;[213] inhibition of lactic acid accumulation by lactic starter cultures in bologna;[123] and, inhibition of toxin release or activation by *C. botulinum*.[218]

4.3. Foods

Testing of phosphates as preservatives for products, such as fruits and fruit juices, egg products, bakery items and cheeses has been limited.[201] Dipping of fresh cherries in solutions of phosphates delayed mold spoilage during refrigerated storage, and extended the period before processing.[134] The longer chain tetraphosphate was a better inhibitor than tripolyphosphate and pyrophosphate. A patent has also described preservation of fruit juices and other foods with long chain polyphosphates.[76]

Another patent deals with pasteurization (52–55°C) of egg whites in the presence of sodium hexametaphosphate (0·5–0·75%). The process destroys *Salmonella* organisms at lower temperatures, without damage to the functional properties of egg proteins, and it also preserves the products during subsequent storage.[35,75,77] Polyphosphates also delay growth of *Staphylococcus* and other Gram positive bacteria in egg whites,[133] and *Bacillus mesentericus* to prevent ropiness in bakery products.[128]

In cheese manufacture, phosphates added to milk inhibit destruction of the starter culture by bacteriophage, since they sequester calcium ions, which are necessary for phage attachment.[85,201] An important anti-microbial contribution of phosphate in dairy products is in processed cheeses, where they inhibit clostridia and other gas-producing organisms, in addition to helping emulsification.[189,190,201]

4.4. Meat Products

Certain phosphates may be beneficial as antimicrobial agents in meat products with reduced levels of sodium chloride. The antimicrobial activity of such formulations, however, has received only limited testing.[175,178,179] The shelf-life of reduced sodium meat formulations is generally shortened and certain phosphates (e.g. sodium acid pyrophosphate) can improve their shelf-life, while others (e.g. sodium tripolyphosphate) are ineffective.[92,93,98,123,124,148,173,175–177,180,225,226]

In general, the antimicrobial activity of phosphates in meat products is

TABLE 10

EFFECT OF SODIUM HEXAMETAPHOSPHATE ON BACTERIAL GROWTH AND HISTAMINE PRODUCTON IN A CULTURE BROTH AFTER 20 HOURS AT 32°C

Bacteria	Sodium hexametaphosphate (%)	Log CFU/g	Histamine concentration (moles/ml)
Proteus morganii—JM	0	8·2	73
	1·0	8·6	56
Klebsiella pneumoniae—T₂	0	6·4	80
	1·0	8·7	73

Data extracted from Taylor and Speckhard,[193] *Journal of Food Protection*, 1984, **47**, 508. Copyright © by International Association of Milk, Food and Environmental Sanitarians.

variable and depends on type of product, type of phosphate, pH, processing variables and other (e.g. sodium chloride, nitrite) inhibitors included in the formulation.[178,201] Under certain conditions, phosphates have delayed microbial growth,[60,62,92,93,120,147,216] while other studies have indicated phosphates as ineffective in microbial inhibition.[61,92,146,176]

Application of phosphate solutions to raw poultry meat by immersion, spraying or injection has reduced bacterial numbers.[15,25,31,80,81,181,184,192] The inhibitory activity, however, depends on several variables. Dipping of fresh chicken parts in a solution of 3% polyphosphate controlled growth of Gram positive micrococci and staphylococci, but certain Gram negative species were resistant to levels of 1–6% phosphate.[15] Viable counts of *Salmonella* in chicken breast meat treated with 0·35% polyphosphate declined faster at −2°C than other (1, −5, −20°C) temperatures.[31] In contrast, phosphate had no effect on *Salmonella* survival when injected in leg muscles. Injecting chicken and other meats with phosphate solutions or other additives may introduce microbial contamination deep into the muscles, and change the composition of the natural microbial flora in the meat.[104] Another study indicated that phosphate may reduce *Salmonella* counts in chicken meat only when used in combination with heat (65–90°C).[199]

Processed meat products tested for microbial inhibition with phosphates include pork slurries,[60,62,146,147] mechanically deboned poultry meat frankfurter batters,[5,120] beef-pork frankfurter batters,[92,93,176,177,216] bologna-type sausages,[20,123] bratwurst,[111,114,115] comminuted poultry meat,[126] beef patties,[116-118,172] and ground pork.[109,172]

Microorganisms examined for sensitivity to phosphates in processed meat products have included total aerobes and anaerobes,[20,92,93,109,111,114-116,176,177,180] lactics,[20,123,124] enterobacteriaceae and enterococci,[20] *Brochothrix thermosphacta*,[123] *Moraxella-Acinetobacter*,[172] *Serratia liquefaciens*,[123] *S. typhimurium*,[126] *S. aureus*,[111,116] *C. sporogenes*[92,93,114,115,176,177] and *C. botulinum*.[5,60-62,120,146,147,216] As indicated above, the effect of phosphates on these microorganisms is variable and depends on several interacting factors.

4.5. Interactions

4.5.1. Phosphate Types and Levels

Microbial inhibition by phosphates varies with type and chain length. Several studies have reported the relative ineffectiveness of orthophosphate, pyrophosphate and tripolyphosphate as inhibitors of microbial growth in meat products.[5,62,92,93,109,111,114,115,120,123,176,177] As a generalization, among alkaline phosphates, the inhibitory activity should increase with increasing chain length. Thus, shorter chain phosphates such as orthophosphate, pyrophosphate and tripolyphosphate are usually ineffective. Specific effects, however, depend on experimental variables and interactions within each system.

Sodium hexametaphosphate has often been found more inhibitory in culture media than other alkaline phosphates,[37,63,152,155] while orthophosphates are usually ineffective.[152,215] At higher concentrations, even tripolyphosphate has been found effective.[182] Tripolyphosphate was more effective than pyrophosphate and orthophosphate against *Moraxella-Acinetobacter* cultures.[30] Pyrophosphate and a cyclic metaphosphate were found ineffective against *S. aureus*.[63] Sodium tetraphosphate was the most inhibitory against mold spoilage of fresh cherries followed by tripolyphosphate and pyrophosphate.[134] Diphosphate was generally more effective than tripolyphosphate in controlling *C. botulinum* growth and toxin production.[62] Tripolyphosphate was ineffective against spoilage of frankfurter batters inoculated with *C. sporogenes*.[92,176,177] In general, longer chain alkaline polyphosphates are more effective as antimicrobial agents than shorter chain compounds. Variability, however, should be expected depending on other factors, such as microbial species, pH, heat processing, phosphate hydrolysis and chemical preservatives.

The most inhibitory polyphosphate appears to be the acidic form of pyrophosphate (i.e. sodium acid pyrophosphate), which is of short chain length (Table 11). Sodium acid pyrophosphate has inhibited mesophilic and psychrotrophic bacteria and *S. aureus* in uncooked bratwurst;[111] aerobes, anaerobes and *C. sporogenes* in cooked bratwurst and frankfurters;[92,93,114,115] *Brochotrix thermosphacta* and *Serratia liquefaciens* in bologna-type sausage;[123] and, *C. botulinum* in poultry and pork–beef frankfurters.[5,120,216] In all these products and against the same microorganisms, other phosphates, such as tripolyphosphate and hexametaphosphate were less effective or noninhibitory. In addition to sodium acid pyrophosphate, two longer chain polyphosphates (12 and 22 average chain length) were more effective than pyrophosphate and tripolyphosphate against growth of spoilage bacteria in frankfurter batters (Table 12).[92] In culture media, however, tetrasodium pyrophosphate was more inhibitory, followed by tripolyphosphate and hexametaphosphate, while sodium acid pyrophosphate was not inhibitory or enhanced recovery.[110]

It appears that reduction in pH caused by sodium acid pyrophosphate may be more critical in the expression of antimicrobial activity than the

TABLE 11
EFFECT OF PHOSPHATES (TRIPOLYPHOSPHATE, TPP; HEXAMETAPHOSPHATE, HMP;
ACID PYROPHOSPHATE, APP) ON TOXIN PRODUCTION BY CLOSTRIDIUM BOTULINUM
IN TURKEY MEAT FRANKFURTERS

NaCl (%)	Phosphate (0·4%)	pH	Days to toxin (27°C)
2·5	—	6·52	4
2·0	TPP	6·71	4
2·0	HMP	6·52	4
2·0	APP	6·30	10

Data extracted from Barbut et al.,[4] Journal of Food Science, 1987, **52**, 1137. Copyright © by Institute of Food Technologists.

longer chain length of alkaline phosphates, which also raise the pH. It should be noticed, however, that sodium acid pyrophosphate inhibits microbial growth more than the inhibition caused by the lower pH values. Thus, the phosphate ion of the compound contributes antimicrobial activity in excess to that caused by reduced pH.[93,178] It will be interesting to determine whether alkaline phosphates show antimicrobial activity at pH values lower than those achieved by their introduction in food systems. One study, however, found sodium acid pyrophosphate more inhibitory than tripolyphosphate and hexametaphosphate at comparable pH values.[120]

Microbial inhibition also varies with commercial blends of phosphates, depending on their composition. In one study with frozen poultry meat, certain commercial blends containing pyrophosphate and tripolyphosphate reduced viability of Salmonella by 70–99% while sodium acid pyrophosphate was the least effective.[126] Thus, the composition of commercial blends should be monitored in order to achieve desirable functional properties and increased product shelf-life.

The influence of increasing phosphate concentrations on microbial growth has not received adequate testing. Although increasing levels may result in greater delays of microbial growth, regulations limit the amounts used in foods based on improvement of functional properties and product quality. Sensitivity of different microbial species is variable with phosphate concentrations. Levels of hexametaphosphate needed to inhibit various species ranged from 0·05%–>10%.[133]

4.5.2. Heat Treatment

Inhibition of microorganisms by phosphates may be affected by heat treatment. Phosphates may also affect the heat resistance and recovery of

TABLE 12

INFLUENCE OF VARIOUS PHOSPHATES, USED AT EQUAL IONIC STRENGTH, ON GAS PRODUCTION BY *CLOSTRIDIUM SPOROGENES* IN FRANKFURTER BATTERS WITH 1·25% SODIUM CHLORIDE STORED AT 27°C

Treatment	Concentration (%)	pH	Days to gas detection
Control	0	6·25	3·5
Sodium acid pyrophosphate	0·17	6·08	8·0
Tetrasodium pyrophosphate	0·20	6·48	4·5
Sodium tripolyphosphate	0·22	6·42	4·0
Sodium tetrametaphosphate	0·28	6·32	3·5
Sodium hexametaphosphate (12[a])	0·33	6·36	5·5
Sodium hexametaphosphate (21[a])	0·34	6·40	5·5

Data extracted from Madril and Sofos,[92] and Sofos and Madril.[180]
[a]Phosphate chain length.

heated microorganisms. Phosphates have also interacted with irradiation to increase the shelf-life of meat products.[4,18,22] A phosphate dipping treatment for chicken carcasses reduced microbial counts only when it was combined with heating.[199] In contrast, filter sterilized phosphate was a more effective inhibitor of colony formation by *Moraxella-Acinetobacter* in culture media than heated (70–121°C) phosphate.[30] Heated or unheated pyrophosphate was highly inhibitory or lethal to *S. typhimurium*, *Pseudomonas fluorescens*, *S. aureus* and lactics in culture media, followed by tripolyphosphate and hexametaphosphate (Table 13).[110] In general, heating reduced the activity of phosphates, with pyrophosphate being the only phosphate retaining its activity upon heating. Sodium acid pyrophosphate enhanced microbial recovery when heated. The antimicrobial activity of tripolyphosphate and hexametaphosphate was lost upon heating, except against the lactic cultures. Tripolyphosphate was less inhibitory against *S. aureus* in culture media after heat sterilization than after filter sterilization, whereas hexametaphosphate retained most of its antimicrobial activity after heating.[63]

Phosphates may affect the heat resistance of microorganisms. Diphos-

TABLE 13
AVERAGE (%) RECOVERY (72 HOURS) OF BACTERIA (24-H AND 3-H CULTURES) IN TRYPTICASE SOY AGAR WITH PHOSPHATE (0·5%)

Phosphate	Heat treatment	Salmonella typhimurium		Staphylococcus aureus		Lactic starter culture	
		24-h	3-h	24-h	3-h	24-h	3-h
TSPP	Heated	49	1	< 1	< 1	< 1	< 1
	Unheated	27	3	< 1	< 1	2	5
STPP	Heated	91	39	89	105	86	54
	Unheated	42	3	< 1	< 1	75	3
SPG	Heated	108	105	87	101	87	41
	Unheated	71	49	< 1	< 1	< 1	< 1
SAPP	Heated	96	134	82	118	108	95
	Unheated	73	87	87	114	116	100

TSPP = tetrasodium pyrophosphate; STPP = sodium tripolyphosphate; SPG = sodium phosphate glassy; SAPP = sodium acid pyrophosphate.
Data extracted from Molins *et al.*,[110] *Journal of Food Science*, 1984, **49**, 948.
Copyright © by Institute of Food Technologists.

phate and tripolyphosphate reduced the heat resistance of *Lactobacillus helveticus, Bacillus cereus* and *C. sporogenes* in sausages, but had no effect on survival of the bacteria in stored products.[26,222] Orthophosphates in sporulation media reduced the heat resistance of *Bacillus* spores.[23] Germination of spores was also reduced by heat treatment in phosphate buffers.[28] In general, phosphates decrease the thermal resistance of several bacteria in foods.[47,75,77,222,232] In contrast, fewer *Moraxella-Acinetobacter* cells were thermally inactivated or injured if sodium chloride and tripolyphosphate were present in beef.[29] Thus, salts, such as sodium chloride and phosphates, may play both a protective and inhibitory effect on microbial destruction.

Autoclaving phosphate in culture media containing glucose has led to formation of stimulatory and/or inhibitory substances.[27] Formation of such compounds in food systems, however, is unknown.[201]

Heat induced injury of salmonellae in culture media was enhanced by polyphosphates, and especially tripolyphosphate.[156] Heat sensitization was more pronounced with increasing heating time and temperature, or higher pH and reduced inoculum concentration. The heat sensitizing effect of polyphosphates was also dependent on the composition of the heating medium and the species of *Salmonella*. Food ingredients had only a slight effect on heat sensitization. Some food ingredients, however, may reduce the efficacy of phosphates through direct interaction or by contributing metal ions to interact with phosphates.

The heat sensitizing effect of phosphates will be useful when there is a need for reduction in the extent of heat treatment to avoid quality defects in pasteurized foods. An example of this is the use of polyphosphates in reducing pasteurization temperatures for liquid egg whites.[75,77] In other foods, however, the results are conflicting. Phosphates reduced *D*-values (i.e. time needed for a given heat treatment to destroy 90% of a microbial population) for spores of *Bacillus licheniformis* in pork, but not in beef paste.[232] Addition of phosphates to milk inoculated with *B. cereus* cells and spores did not sensitize the spores to heat and resulted in a shortened shelf-life.[157] The higher heat sensitivity of bacteria in presence of phosphates may be due to chelation of metal ions that stabilize membranes or are involved in reactions as cofactors.

Phosphates may be more inhibitory to recovery of heat-stressed microorganisms. Heating of cultures increased the inhibitory effect of commercial polyphosphate blends against *S. aureus, S. faecalis, B. subtilis* and *C. bifermentans*.[70]

4.5.3. Hydrolysis

Hydrolysis of polyphosphates to shorter chain compounds and eventually to orthophosphate will influence their antimicrobial activity, because there is variation in antimicrobial activity among phosphate types. It is believed that phosphates added to uncooked muscle foods are hydrolyzed by naturally present phosphatase enzymes.[3,109,111,112,114,115,121,185,194] Thus, the length of time between addition of polyphosphates and heat treatment can influence the effect of polyphosphates on microbial growth.

Cooking of bratwurst helped retain the antimicrobial properties of phosphates.[111,115] Heat treatment, however, may also result in hydrolytic degradation of phosphates.[30] At 70°C and 121°C in pH 7·0–7·5 aqueous media hydrolysis of half of the tripolyphosphate to orthophosphate required 50 hours, and 21 minutes, respectively.[209] Since orthophosphate is ineffective and longer chain polyphosphates are usually more effective, filter sterilized phosphate should be more inhibitory than autoclaved.[30] When enzyme hydrolysis is involved, however, heating of the food immediately after formulation should increase antimicrobial activity. Certain phosphates (e.g. pyrophosphate), however, have shown stability to hydrolysis by heat.[6]

In addition to phosphate type, heat treatment and presence of enzymes, hydrolysis of phosphates and microbial inhibition may also be influenced by pH and product composition. Interactions of phosphates with meat proteins other than enzymes may also affect their antimicrobial activity.[122,125] Hydrolysis of polyphosphates does not explain the reportedly higher antimicrobial activity of longer chain compounds.[201] It is difficult, however, to determine the extent of hydrolysis in muscle foods and its correlation with microbial inhibition.[109,112-114]

4.5.4. Product pH

The antimicrobial activity of phosphates, like other food preservatives, should be affected by the pH of the substrate. Product pH will affect not only microbial growth, but also reactivity of phosphates. There is a need, however, for more reseach to examine the relationship between pH and the antimicrobial activity of phosphates.

Although some studies have reported increased antimicrobial activity at higher (> 7·4) pH values,[63] others have indicated that the activity is not due to an increase in pH.[25,30] There is a need, however, for further examination of the antimicrobial activity of alkaline phosphates at both lower and higher pH values.[178] A polyphosphate blend increased the probability of toxin production by C. botulinum in a pork slurry at pH values of 5·5–6·3,[146] but in some instances it decreased it at the higher pH values of 6·3–6·8 (Table 14).[147] A long chain (22 phosphate units) polyphosphate, however, was more inhibitory against *C. botulinum* in culture media of pH 5·4–6·0 than 6·6.[152] Sodium hexametaphosphate was also more inhibitory of *Bacillus* spores at pH 6·0 than 7·0.[37]

The antimicrobial activity of sodium acid pyrophosphate is generally higher at lower pH values.[93,120,217,219] Inhibition of microbial growth by the compound in frankfurter batters, however, was due not only to lower pH values, but also due to the phosphate ion which provided additional inhibition. The data also suggested that the compound may be a better inhibitor at pH 6·0 than 5·7 and 6·3. This observation needs verification, because it may be related to the mechanism of antimicrobial activity of phosphates.[93,178]

4.5.5. Microbial Species

Microbial inhibition by phosphates varies with types and species of microorganisms. Gram positive bacteria are more sensitive to inhibition by phosphates than Gram negative species.[15,25,110,133,191] Fluorescent strains of pseudomonads are more resistant to inhibition than nonfluorescent strains.[25] *Salmonella typhimurium* and *Pseudomonas aeruginosa* are less sensitive to inhibition than *S. aureus* and lactic acid starter cultures.[110] A mixture of pyrophosphate and orthophosphate did not inhibit aerobic mesophiles and lactics, and stimulated enterobacteriaceae in bologna-type sausages.[20] In a similar product, tripolyphosphate was ineffective against *B. thermosphacta* and *S. liquefaciens*, while an acidic phosphate was inhibitory.[123] Sensitivity of microbial cultures, however, is often dependent on the age of the cells. Twenty-four-hour cultures were more resistant to inhibition by phosphates than 3-hour cultures (Table 13).[110]

4.5.6. Chemical Additives

The antimicrobial activity of phosphates in a given substrate may also depend on interactions with other chemical additives present.[178] The common food preservative sodium chloride had additive activity with pyrophosphate and synergistic activity with tripolyphosphate against *Moraxella-Acinetobacter* in culture media.[30] Sodium acid pyrophosphate, however, showed no significant interaction with sodium chloride against *C. botulinum* in culture media.[217,219] Actually presence of sodium chloride appeared to decrease the antimicrobial activity of sodium acid pyrophosphate–sorbate combinations. Elimination of sodium chloride from the formulation increased the antimicrobial activity of phosphate–sorbate, but

JOHN N. SOFOS

TABLE 14

INFLUENCE OF SODIUM CHLORIDE, pH AND A PHOSPHATE BLEND (CURAPHOS 700) ON THE PROBABILITY (%) OF TOXIN PRODUCTION BY *CLOSTRIDIUM BOTULINUM* IN PORK SLURRIES (15°C)

NaCl (%, w/v)	pH 5·5–6·3		pH 6·3–6·8	
	No phosphate	*0·3% phosphate*	*No phosphate*	*0·3% phosphate*
2·5	48	77	81	53
3·5	14	28	66	33
4·5	3	4	29	9

Data extracted from Robinson *et al.*,[148] *Journal of Food Technology*, 1982, **17** 727. Copyright © Blackwell Scientific Publications.

reduction of sodium chloride from 2·5% to 1·25% (in the water phase) reduced inhibition by phosphate–sorbate. At these low levels, however, sodium chloride may have acted as a protective or stimulating agent on bacterial spores.[178] In a pork slurry, inhibition of *C. botulinum* by phosphate increased with increasing concentration of sodium chloride (Table 14).

The increased inhibition by sodium chloride–phosphate combinations does not appear to be due to lower water activity values.[172] There was strong inhibition of *S. aureus* in phosphate buffer when the water activity was reduced to 0·90 with sodium or potassium chloride. No such inhibition existed, however, when sucrose, glucose or glycerol were used to reduce the water activity.[208] Phosphate levels used in foods have no detectable effect on the water activity.[126]

Phosphates have also interacted with nitrite and sorbate in culture media and in foods.[34,178] In culture media, tripolyphosphate in combination with sorbate resulted in *C. botulinum* cells of abnormal shape and defective cell division.[155] Similar interactions were observed with combinations of sodium acid pyrophosphate and sorbate against *C. botulinum*[217,219] and *S. aureus*.[198] Combinations of nitrite, sorbate and sodium acid pyrophosphate have also increased microbial inhibition in meat products.[60,114,120,216]

4.6. Mechanisms

Similar to several other food additives, the mechanism of antimicrobial action by phosphates is largely unknown.[178,201,215] Several potential modes of action, however, have been reported and will be discussed.

The inability of phosphates to inhibit microbial growth under certain conditions appears to be due to variations in factors such as microbial species; phosphate type and level; heat treatment and phosphate hydrolysis; composition and pH of the substrate; chemical additives; and storage conditions.[178] The interaction of these factors also complicates attempts to determine modes and mechanisms of action.

Possible mechanisms of action by phosphates may include interaction with cell walls and membranes; inhibition of enzymes or transport functions; or other unknown types of action. These activities may be influenced by changes in pH, ionic strength or chelation of metal cations by phosphates.

Increases in ionic strength have been reported as important in increases of product binding with phosphates.[204,205] The influence of ionic strength on microbial growth, however, is not well defined. Generally, changes in pH and ionic strength may have some effect on the antimicrobial activity of phosphates, but they cannot be used to completely understand modes and mechanisms of action.[25,30,215] Ionic strength, and, especially, pH can affect reactivity of phosphates with food and cell components, and thus, influence modes of action under different conditions.

Several studies have reported that phosphates inhibited spore germination at certain pH values (5·55) or cell division at others (pH 5·85);[220] or contributed to elongation of cells and abnormal cell division.[37,155,215,218,220] Elongation of the cells may have been caused by interaction of phosphate with inner protein components of the cell, which may have increased water binding within the cell structure.[215] This may have increased cell size directly, or modified normal cell development. Other, unknown, metabolic modifications by phosphates may also have occurred.

Indications exist that the ability of phosphates to chelate cations may be essential in their antimicrobial properties,[24,25,35,133,178,201,215] Lysis of bacteria by hexametaphosphate was prevented by sodium chloride and magnesium sulfate, which also permitted growth.[133] This suggests that hexametaphosphate chelated metal cations (e.g. magnesium) and interfered with metabolism, which inhibited cell division and caused loss of cell wall integrity. The higher requirement for magnesium by Gram negative bacteria also explains their increased sensitivity to phosphate compared to Gram positive species.[133] This mode of action is supported by studies indicating that addition of magnesium and competitive chelators (e.g. pyoverdine and peptone) reversed inhibition by phosphate.[25]

Supplementation of broth with magnesium was also effective in overcoming inhibition of S. aureus by phosphate.[63] Inhibition was partially reversed by calcium and iron but not by zinc and manganese. Metal (e.g. iron) chelation by sodium acid pyrophosphate was also presented as a reason to explain synergistic antibotulinal effects of the compound with sorbate in the presence, but not in the absence, of nitrite.[120]

Metal ion chelation is affected by phosphate type and pH. Additional studies should be designed to determine if the chelating ability of various phosphates at different pH values agrees with their antimicrobial activity. Chain polyphosphates are stronger chelators than cyclic phosphates, while orthophosphates do not chelate.[58,59,210,212] Long chain polyphosphates (e.g. hexametaphosphate) are strong chelators of calcium and magnesium and more effective at higher pH values.[161] Shorter chain polyphosphates (e.g. pyrophosphate) are better chelators of iron and copper, and their chelating capacity decreases as the pH increases.[183]

It should be noted that chelators may either enhance or hinder microbial growth.[178] Metals essential for growth may be more available after chelation by assimilable compounds.[184] Chelators may also remove toxic metals from substrates,[56,103] or inhibit metabolism of important nutrients.[84,103] Thus, it is difficult to estimate the importance of metal ion chelation in antimicrobial activity by phosphates.

Inhibition of enzymes and subsequent inhibition of transport function or metabolic activity may also be involved in the antimicrobial action by phosphates.[178,215] Inhibition of enzymes such as 5-nucleotidase and protease[215,218] by phosphates (e.g. sodium acid pyrophosphate) has been reported. Sodium acid pyrophosphate delayed toxicity of *C. botulinum* cultures.[218,219] The nontoxic proteolytic cultures, however, became toxic when treated with trypsin. The results suggested that sodium acid pyrophosphate inhibited production or function of protease enzymes involved in toxin activation.[215] It was speculated that such inhibition could be through binding of the phosphate to protease molecules or chelation of metal cations that may be needed for enzyme activation.[218,219] A subsequent study,[221] using radioactively labeled phosphates, indicated that orthophosphate and sodium acid pyrophosphate were associated with vegetative cells during growth. Orthophosphate, however, was released at the stationary phase or during cell lysis, while acid pyrophosphate was not released. More of the pyrophosphate was associated with cell RNA than DNA. This can explain the lack of microbial inhibition by orthophosphate, and indicates that synthesis of proteases produced from RNA may be inhibited by sodium acid pyrophosphate.[221] More studies, however, are needed to determine exact mechanisms of enzyme inhibition by phosphates.

REFERENCES

1. ACTON, J. C., ZIEGLER, G. R. and BURGE, D. L. *CRC Crit. Rev. Food Sci. Nutr.*, 1983, **18**, 99.
2. AHMED, E. M., CORNELL, J. A., TOMASZEWSKI, F. B. and DENG, J. C., *J. Food Sci.*, 1983, **48**, 1078.
3. AWAD, M. K. M.Sc. thesis, 1968, University of Alberta, Edmonton.
4. BARBUT, S., MAURER, A. J. and THAYER, D. W. *J. Food Sci.*, 1987, **52**, 1137.
5. BARBUT, S., TANAKA, N., CASSENS, R. G. and MAURER, A. J. *J. Food Sci.*, 1986, **51**, 1136.
6. BELL, R. N. *Ind. Eng. Chem.*, 1947, **39**, 136.
7. BENDALL, J. R. *J. Sci. Food Agric.*, 1954, **5**, 468.
8. BICKEL, W. 1956, US Patent 2 735 776.
9. BROTSKY, E. *Poultry Sci.*, 1976, **55**, 653.
10. BROTSKY, E. and EVERSON, C. W. *Proc. Meat Ind. Res. Conf. Amer. Meat Inst.*, Washington, DC, 1973, p. 107.

11. BURGIN, J. M., ROGERS, R. W. and AMMERMAN, G. R. *J. Food Sci.*, 1985, **50**, 1598.
12. CARPENTER, J. A., SAFFLE, R. L. and KAMSTRA, L. D. *Food Technol.*, 1961, **15**, 197.
13. CASSIDY, J. P. *Food Prod. Dev.*, 1977, **11**, 74.
14. CHANG, I. and WATTS, B. M. *Food Technol.*, 1949, **3**, 332.
15. CHEN, T. C., CULOTTA, J. T. and WANG, W. S. *J. Food Sci.*, 1973, **38**, 155.
16. CHU, Y. H., HUFFMAN, D. L., TROUT, G. R. and EGBERT, W. R. *J. Food Sci.*, 1987, **52**, 869.
17. CLARKE, A. D., MEANS, W. J. and SCHMIDT, G. R. *J. Food Sci.*, 1987, **52**, 854.
18. COHEN, J. S., SHULTS, G. W., MASON, V. C. and WIERBICKI, E. *J. Food Sci.*, 1977, **42**, 338.
19. COREY, M. L., GERDES, D. L. and GRODNER, R. M. *J. Food Sci.*, 1987, **52**, 297.
20. DAELMAN, W. and VAN HOOF, J. *Arch. Lebensmittelhyg.*, 1975, **26**, 213.
21. DEMAN, J. M. and MELNYCHYN, P. *Symp.: Phosphates in Food Processing*, AVI Publishing Co., Westport, 1971.
22. EHIOBA, R. M., KRAFT, A. A., MOLINS, R. A., WALKER, H. W., OLSON, D. G., SUBBARAMAN, G. and SKOWRONSKI, R. P. *J. Food Sci.*, 1987, **52**, 1477.
23. EL-BISHI, H. M. and ORDAL, Z. J. *J. Bacteriol.*, 1956, **71**, 1.
24. ELLINGER, R. H. *Phosphates as Food Ingredients*, CRC Press, Cleveland, 1972.
25. ELLIOT, R. P., STRAKA, R. P. and GARIBALDI, J. A. *Appl. Microbiol.*, 1964, **12**, 517.
26. FAHNLE, H., WATSOS, E., STAUSS, H., OZARI, R., SCHMIDT, H. and KOTTER, L. *Fleischwirtschaft*, 1985, **65**, 485.
27. FINKELSTEIN, R. A. and LANKFORD, C. E. *Appl. Microbiol.*, 1957, **5**, 74.
28. FINLEY, N. and FIELDS, M. L. *Appl. Microbiol.*, 1962, **10**, 231.
29. FIRSTENBERG-EDEN, R., ROWLEY, D. B. and SHATTUCK, G. E. *Appl. Environ. Microbiol.*, 1980, **40**, 480.
30. FIRSTENBERG-EDEN, R., ROWLEY, D. B. and SHATTUCK, G. E. *J. Food Sci.*, 1981, **46**, 579.
31. FOSTER, R. D. and MEAD, G. C. *J. Appl. Bacteriol.*, 1976, **41**, 505.
32. FRONING, G. W. and SACKETT, B. *Poultry Sci.*, 1985, **64**, 1328.
33. FUKAZAWA, T., HASHIMOTO, Y. and YASUI, T. *J. Food Sci.*, 1961, **26**, 550.
34. FUKUI, T., MORI, K., SAWADA, G. and AKABANE, Y. 1974, Japanese Patent 74 46 070.
35. GARIBALDI, J. A., IJICHI, K. and BAYNE, H. G. *Appl. Microbiol.*, 1969, **18**, 318.
36. GILLETT, T. A., MEIBURG, D. E., BROWN, C. L. and SIMON, S. *J. Food Sci.*, 1977, **42**, 1606.
37. GOULD, G. W. *Microbial Inhibitors in Food* (ed. N. Molin), Almqvist and Wiksell Stockholm, Sweden, 1964, p. 17.
38. GOVINDARAJAN, S., HULTIN, H. O. and KOTULA, A. W. *J. Food Sci.*, 1977, **42**, 571.
39. GRAY, G. W. and WILKINSON, S. G. *J. Appl. Bacteriol.*, 1965, **28**, 153.
40. HALLIDAY, D. A. *Process Biochem.*, 1978, **13**, 6.
41. HAMM, R. *Adv. Food Res.*, 1960, **10**, 355.
42. HAMM, R. *Symp. on Phosphates in Food Proc.* (eds J. M. Deman and P. Melnychyn), AVI Publishing Co., Westport, 1970, p. 65.
43. HAMM, R. *Deutsche Lebensmittel-Rundschau*, 1980, **76**, 263.

44. HAMM, R. and GRAU, R. *Z. Lebensm. Untersuch. U-Forsch.*, 1958, **108**, 280.
45. HAMM, R. and NERAAL, R. *Fleisch. Z. Lebensm. Unteforsch.-Forsch.*, 1977, **164**, 243.
46. HARGETT, S. M., BLUMER, T. N., HAMMAN, D. D., KEETON, J. T. and MONROE, R. J. *J. Food Sci.*, 1980, **45**, 905.
47. HARGREAVES, L. L., WOOD, J. M. and JARVIS, B. Scientific and Technical Survey No. 76, British Food Manufacturing Industries Research Association, Leatherhead, UK, 1972.
48. HAROLD, F. M. *Bacteriol. Rev.*, 1966, **30**, 772.
49. HAYMON, L. W., BROTSKY, E., DANNER, W. E., EVERSON, C. W. and HAMMER, P. A. *J. Food Sci.*, 1976, **41**, 417.
50. HELLENDOORN, B. W. *Food Technol.*, 1962, **16**, 119.
51. HOES, T. L., RAMSEY, C. B., HINES, R. C. and TATUM, J. D. *J. Food Sci.*, 1980, **45**, 773.
52. HOPKINS, E. W. and ZIMONT, L. J., 1957, US Patent 2 999 019.
53. HUFFMAN, D. L., ANDE, C. F., CORDRAY, J. C., STANLEY, M. H. and EGBERT, W. R. *J. Food Sci.*, 1987, **52**, 275.
54. HUFFMAN, D. L., CROSS, H. R., CAMPBELL, K. J. and CORDRAY, J. C. *J. Food Sci.*, 1981, **46**, 34.
55. HUTCHINS, B. K., LIU, T. H. P. and WATTS, B. M. *J. Food Sci.*, 1967, **32**, 214.
56. HUTNER, S. H., PROVASOLI, L., SCHATZ, A. and HASKINS, C. P. *Proc. Amer. Phil. Soc.*, 1950, **94**, 152.
57. INKLAAR, P. A. *J. Food Sci.*, 1967, **32**, 525.
58. IRANI, R. R. and CALLIS, C. F. *J. Amer. Oil Chem. Soc.*, 1962, **39**, 156.
59. IRANI, R. R. and MORGENTHALER, W. W. *J. Amer. Oil Chem. Soc.*, 1963, **40**, 283.
60. IVEY, F. J. and ROBACH, M. C. *J. Food Sci.*, 1978, **43**, 1782.
61. IVEY, F. J., SHAVER, K. J., CHRISTIANSEN, L. N. and TOMPKIN, R. B. *J. Food Prot.*, 1978, **41**, 621.
62. JARVIS, B., RHODES, A. C. and PATEL, M. *Proc. Int. Meet. Food Microbiol. Technol.* (eds B. Jarvis, J. A. B. Christian and H. D. Michener), Medicina Viva Servizio Congressi, Parma, Italy, 1979, p. 251.
63. JEN, C. M. C. and SHELEF, L. A. *Appl. Environ. Microbiol.*, 1986, **52**, 842.
64. JONES, J. M., GRIFFITHS, N. M., GREY, T. C., WILKINSON, C. C. and ROBINSON, D. *Lebensm.-Wiss. u.-Technol.*, 1980, **13**, 145.
65. JONES, S. L., CARR, T. R. and MCKEITH, F. K. *J. Food Sci.*, 1987, **52**, 279.
66. KAMSTRA, L. D. and SAFFLE, R. L. *Food Technol.*, 1959, **13**, 652.
67. KARMAS, E. *Meat Product Manufacture*, Noyes Data Corporation, Park Ridge, NJ, 1970, p. 38.
68. KEETON, J. *J. Food Sci.*, 1983, **48**, 878.
69. KEETON, J. T., FOEGEDING, E. A. and PATANA-ANAKE, C. *J. Food Sci.*, 1984, **49**, 1462.
70. KELCH, F. and BUHLMANN, X. *Fleishwirt.*, 1958, **10**, 325.
71. KIM, D., KIM, Y., KIM, I. and LEE, B. *Korean J. Food Sci. Technol.*, 1985, **17**, 253.
72. KNIPE, C. L., OLSON, D. G. and RUST, R. E. *J. Food Sci.*, 1985, **50**, 1014.
73. KNIPE, C. L., OLSON, D. G. and RUST, R. E. *J. Food Sci.*, 1985, **50**, 1010.
74. KNIPE, C. L., OLSON, D. G. and RUST, R. E. *J. Food Sci.*, 1985, **50**, 1017.

75. KOHL, W. F. *Food Technol.*, 1971, **25**, 1176.
76. KOHL, W. F. and ELLINGER, R. H. 1972, US Patent 3 681 091.
77. KOHL, W. F., SOURBY, J. C. and ELLINGER, R. H. 1970, US Patent 3 520 700.
78. KONDAIAH, N., ANJANEYULU, A. S. R., RAO, V. K., SHARMA, N. and JOSHI, H. B. *Meat Sci.*, 1985, **15**, 183.
79. KOTTER, L. *Fleischwirtschaft*, 1960, **13**, 186.
80. KRAFT, A. A. and AYRES, J. C. *Appl. Microbiol.*, 1961, **9**, 549.
81. KRAFT, A. A. and AYRES, J. C. *J. Food Sci.*, 1964, **29**, 218.
82. KRAUSE, R. J., OCKERMAN, H. W., KROL, B., MOERMAN, P. C. and PLIMPTON, R. F., Jr. *J. Food Sci.*, 1978, **43**, 853.
83. LAMKEY, J. W., MANDIGO, R. W. and CALKINS, C. R. *J. Food Sci.*, 1986, **51**, 873.
84. LANKFORD, C. E., KUSTOFF, T. Y. and SERGEANT, R. P. *J. Bacteriol.*, 1957, **74**, 737.
85. LAWRENCE, R. C., THOMAS, T. C. and TERZAGHI, B. E. *J. Dairy Res.*, 1976, **43**, 141.
86. LEWIS, D. F., GROVES, K. H. M. and HOLGATE, J. H. *Food Microstructure*, 1986, **5**, 53.
87. LIU, H. P. *J. Food Sci.*, 1970, **35**, 590.
88. LIU, H. P. *J. Food Sci.*, 1970, **35**, 593.
89. LIVINGSTON, D. J. and BROWN, W. D. *Food Technol.*, 1981, **35**, 244.
90. LYON, C. E. *Poultry Sci.*, 1980, **59**, 1031.
91. LYON, B. G. *Poultry Sci.*, 1983, **62**, 321.
92. MADRIL, M. T. and SOFOS, J. N. *Lebensm.-Wiss. u.-Technol.*, 1985, **18**, 316.
93. MADRIL, M. T. and SOFOS, J. N. *J. Food Sci.*, 1986, **51**, 1147.
94. MAHON, J. B. *Proc. Meat Ind. Res. Conf. Amer. Meat Inst.*, Washington, DC, 1961, p. 59.
95. MAHON, J. H. 1962, US Patent 3 036 923.
96. MAHON, J. H. 1963, US Patent 3 104 170.
97. MAHON, J. H., SCHLAMB, K. and BROTSKY, E. *Phosphates in Food Processing Symposium* (eds J. M. Deman and P. Melnychyn), AVI Publishing Co., Westport, 1970, p. 158.
98. MARCY, J. A., KRAFT, A. A., OLSON, D. G., WALKER, H. W. and HOTCHKISS, D. K. *J. Food Sci.*, 1985, **50**, 316.
99. MARION, W. W. and FORSYTHE, R. H. *J. Food Sci.*, 1964, **29**, 530.
100. MATLOCK, R. G., TERRELL, R. N., SAVELL, L. W., RHEE, K. S. and DUTSON, T. R. *J. Food Sci.*, 1984, **49**, 1363.
101. MATLOCK, R. G., TERRELL, R. N., SAVELL, L. W., RHEE, K. S. and DUTSON, T. R. *J. Food Sci.*, 1984, **49**, 1372.
102. MAWSON, R. F. and SCHMIDT, G. R. *J. Food Sci.*, 1983, **48**, 1705.
103. MAYER, G. D. and TRAXLER, R. W. *J. Bacteriol.* 1962, **83**, 1281.
104. MEAD, G. C. and ADAMS, B. W. *J. Hyg.*, 1979, **82**, 133.
105. MEYER, A. 1956, US Patent 2 735 777.
106. MIHALYI-KENGYEL, V. and KORMENDY, L. *Acta Aliment.* 1973, **2**, 69.
107. MILLER, E. M. and HARRISON, D. L. *Food Technol.* 1965, **19**, 94.
108. MILLER, M. F., DAVIS, G. W., SEIDEMAN, S. C., RAMSEY, C. B. and ROLAN, T. L. *J. Food Sci.*, 1986, **51**, 1435.
109. MOLINS, R. A., KRAFT, A. A. and MARCY, J. A. *J. Food Sci.*, 1987, **52**, 513.

110. MOLINS, R. A., KRAFT, A. A., OLSON, D. G. and HOTCHKISS, D. K. *J. Food Sci.*, 1984, **49**, 948.
111. MOLINS, R. A., KRAFT, A. A. and OLSON, D. G. *J. Food Sci.*, 1985, **50**, 531.
112. MOLINS, R. A., KRAFT, A. A. and OLSON, D. G. *J. Food Sci.*, 1985, **50**, 1482.
113. MOLINS, R. A., KRAFT, A. A. and OLSON, D. G. *J. Food Sci.*, 1987, **52**, 1486.
114. MOLINS, R. A., KRAFT, A. A., OLSON, D. G., WALKER, H. W. and HOTCHKISS, D. K. *J. Food Sci.*, 1986, **51**, 726.
115. MOLINS, R. A., KRAFT, A. A., WALKER, H. W. and OLSON, D. G. *J. Food Sci.*, 1985, **50**, 876.
116. MOLINS, R. A., KRAFT, A. A., WALKER, H. W., RUST, R. E., OLSON, D. G. and MERKENICH, K. *J. Food Sci.*, 1987, **52**, 46.
117. MOLINS, R. A., KRAFT, A. A., WALKER, H. W., RUST, R. E., OLSON, D. G. and MERKENICH, K. *J. Food Sci.*, 1987 **52**, 50.
118. MOLINS, R. A., SANDOVAL, A. E., OLSON, D. G., RUST, R. E. and KNIPE, C. L. *J. Food Sci.*, 1987, **52**, 851.
119. NAKHOST, Z. and KAREL, M. *J. Food Sci.*, 1985, **50**, 1748.
120. NELSON, K. A., BUSTA, F. F., SOFOS, J. N. and WAGNER, M. K. *J. Food Prot.*, 1983, **46**, 846.
121. NERAAL, F. and HAMM, R. *19th Europ. Meet. Meat Res. Work*, Paris, France, 1973.
122. NEWBOLD, R. P. and TUME, R. K. *J. Food Sci.*, 1981, **46**, 1327.
123. NIELSEN, H.-J. S. and ZEUTHEN, P. *J. Food Prot.*, 1983, **46**, 1078.
124. NIELSEN, H.-J. S. and ZEUTHEN, P. *J. Food Prot.*, 1985, **48**, 150.
125. NIKKILA, O. E., KUUSI, T. and KYTOKANGAS, R. *J. Food Sci.*, 1967, **32**, 686.
126. OBAFEMI, A. and DAVIES, R. *Food Chem.*, 1985, **18**, 179.
127. OFFER, G. and TRINICK, J. *Meat Sci.*, 1983, **8**, 245.
128. O'LEARY, D. K. and KRALOVEC, R. D. *Cereal Chem.*, 1941, **18**, 730.
129. PARK, J. W. and LANIER, T. C. *J. Food Sci.*, 1987, **52**, 1509.
130. PEPPER, F. H. and SCHMIDT, G. R. *J. Food Sci.*, 1975, **40**, 227.
131. PERCEL, P. J., PARRETT, N. A., PLIMPTON, R. F., OCKERMAN, H. W., KROL, B. and VAN ROON, P. S. *J. Food Sci.*, 1982, **47**, 359.
132. PETERSON, D. W. *J. Food Sci.*, 1977, **42**, 100.
133. POST, F. J., KRISHNAMURTY, G. B. and FLANAGAN, M. D. *Appl. Microbiol.*, 1963, **11**, 430.
134. POST, F. J., COBLENTZ, W. S., CHOU, T. W. and SALUNKHE, D. K. *Appl. Microbiol.*, 1968, **16**, 138.
135. PRUSA, K. J. and BOWERS, J. A. *J. Food Sci.*, 1984, **49**, 709.
136. PRUSA, K. J., BOWERS, J. A. and CRAIG, J. A. *J. Food Sci.*, 1984, **49**, 968.
137. PRUSA, K. J., BOWERS, J. A. and CRAIG, J. A. *J. Food Sci.*, 1985, **50**, 573.
138. PUOLANNE, E. J. and MATIKKALA, M. *Fleischwirtsch.*, 1980, **60**, 1233.
139. PUOLANNE, E. J. and RUUSUNEN, M. *Fleischwirtsch.*, 1980, **60**, 1359.
140. PUOLANNE, E. J. and TERRELL, R. N. *J. Food Sci.*, 1983, **48**, 1022.
141. PUOLANNE, E. J. and TERRELL, R. N. *J. Food Sci.*, 1983, **48**, 1036.
142. RAINESBELL, R., DRAPER, H. H., TZENG, D. Y. M., SHIN, H. K. and SCHMIDT, G. R. *J. Nutr.*, 1977, **107**, 42.
143. RAMSEY, M. B. and WATTS, B. M. *Food Technol.*, 1963, **17**, 1056.
144. RANKEN, M. D. *Chem. Ind.*, 1976, **24**, 1052.
145. RHEE, K. S. and SMITH, G. C. *J. Food Prot.*, 1984, **47**, 182.

250 JOHN N. SOFOS

146. ROBERTS, T. A., GIBSON, A. M. and ROBINSON, A. *J. Food Technol.*, 1981, **16**, 239.
147. ROBERTS, T. A., GIBSON, A. M. and ROBINSON, A. *J. Food Technol.*, 1981, **16**, 267.
148. ROBINSON, A., GIBSON, A. M. and ROBERTS, T. A. *J. Food Technol.*, 1982, **17**, 727.
149. SAIR, L. and KOMARIK, S. L. 1968, US Patent 3 391 007.
150. SCHMIDT, G. R., MAWSON, R. F. and SIEGEL, D. G. *Food Technol.*, 1981, **35**, 235.
151. SCHMIDT, G. R. and TROUT, G. R. *Proc. Intl. Symp. Meat Sci. Technol.* (eds K. R. Franklin and H. R. Cross), National Live Stock and Meat Board, Chicago, 1982, p. 265.
152. SCHOENI, J. L., DOYLE, M. and TANAKA, N. *Annual Report Food Reseach Institute, University of Wisconsin*, Madison, WI, 1980, p. 337.
153. SCHWARTZ, W. C. and MANDIGO, R. W. *J. Food Sci.*, 1976, **41**, 1266.
154. SEMAN, D. L., OLSON, D. G. and MANDIGO, R. W. *J. Food Sci.*, 1980, **45**, 1116.
155. SEWARD, R. A., DEIBEL, R. H. and LINDSAY, R. C. *Appl. Environ. Microbiol.*, 1982, **44**, 1212.
156. SEWARD, R. A., LIN, C. F. and MELACHOURIS, N. *J. Food Sci.*, 1986, **51**, 471.
157. SHEHATA, T. E. and COLLINS, E. B. *J. Dairy Sci.*, 1972, **55**, 1405.
158. SHERMAN, P. *Food Technol.*, 1961, **15**, 79.
159. SHIMP, L. A. *Meat Process.*, August, 1981, p. 65.
160. SHIMP, L. A. *Food Engin.*, September, 1983, p. 106.
161. SHIMP, L. A. *Meat Indust.*, November, 1983, p. 24.
162. SHULTS, G. W., RUSSELL, D. R. and WIERBICKI, E. *J. Food Sci.*, 1972, **37**, 860.
163. SHULTS, G. W. and WIERBICKI, E. *J. Food Sci.*, 1972, **37**, 860.
164. SHULTS, G. W. and WIERBICKI, E. *J. Food Sci.*, 1973, **38**, 991.
165. SIEGEL, D. G., CHURCH, K. E. and SCHMIDT, G. R. *J. Food Sci.*, 1979, **44**, 1276.
166. SIEGEL, D. G. and SCHMIDT, G. R. *J. Food Sci.*, 1979, **44**, 1686.
167. SIEGEL, D. G., THENO, D. M. and SCHMIDT, G. R. *J. Food Sci.*, 1978, **43**, 327.
168. SIEGEL, D. G., THENO, D. M., SCHMIDT, G. R. and NORTON, H. W. *J. Food Sci.*, 1978, **43**, 331.
169. SMITH, L. A., SIMMONS, S. L., MCKEITH, F. K., BECHTEL, P. J. and BRADY, P. L. *J. Food Sci.*, 1984, **49**, 1636.
170. SMITH, M. L. and BOWERS, J. A. *Poultry Sci.*, 1972, **51**, 998.
171. SNEDEKER, S. M., SMITH, S. A. and GREGER, J. L. *J. Nutr.*, 1982, **112**, 136.
172. SNYDER, L. D. and MAXCY, R. B. *J. Food Sci.*, 1979, **44**, 33.
173. SOFOS, J. N. *J. Food Sci.*, 1983, **48**, 1684.
174. SOFOS, J. N. *J. Food Sci.*, 1983, **48**, 1692.
175. SOFOS, J. N. *J. Food Safety*, 1984, **6**, 45.
176. SOFOS, J. N. *J. Food Sci.*, 1985, **50**, 1379.
177. SOFOS, J. N. *Proc. Europ. Meet. Meat Res. Work.*, 1985, **31**, 594.
178. SOFOS, J. N. *Food Technol.*, 1986, **40**, 52.
179. SOFOS, J. N. *The Shelf-life of Foods and Beverages* (ed. G. Charalambous), Elsevier, Amsterdam, 1986, p. 463.
180. SOFOS, J. N. and MADRIL, M. T. *Proc. 32nd Europ. Meet. Meat Res. Work.*, Vol. 2, Ghent, Belgium, 1986, p. 341.
181. SPENCER, J. V. and SMITH, L. E. *Poultry Sci.*, 1962, **41**, 1685.

182. STAUFFER CHEMICAL CO. 1969, British Patent 1 154 079.
183. STEINHAUER, J. E. *Dairy Food Sanitat.*, 1983, **3**, 244.
184. STEINHAUER, J. E. and BANWART, G. J. *Poultry Sci.*, 1964, **43**, 618.
185. SUTTON, A. H. *J. Food Technol.*, 1973, **8**, 185.
186. SWIFT, C. E. and ELLIS, R. *Food Technol.*, 1956, **10**, 546.
187. SWIFT, C. E. and ELLIS, R. *Food Technol.*, 1957, **11**, 450.
188. SZTEYN, J. *Medycyna Weterynaryjna*, 1982, **38**, 13.
189. TANAKA, N. *J. Food Prot.*, 1982, **45**, 1044.
190. TANAKA, N., GOEPFERT, J. M., TRAISMAN, E. and HOFFBECK, W. M. *J. Food Prot.*, 1979, **42**, 787.
191. TATSUGUCHI, K. and WATANABE, T. *J. Food Hyg. Soc. Japan*, 1983, **24**, 469.
192. TAYLOR, M. H., SMITH, L. T. and MITCHELL, J. D. *Poultry Sci.*, 1965, **44**, 297.
193. TAYLOR, S. L. and SPECKHARD, M. W. *J. Food Prot.*, 1984, **47**, 508.
194. TENHET, V., FINNE, G., NICKELSON II, R. and TOLODAY, D. *J. Food Sci.*, 1981, **46**, 344.
195. TERRELL, R. N., SWASDEE, R. L., SMITH, G. C., HEILIGMAN, F., WIERBICKI, E. and CARPENTER, Z. L. *J. Food Prot.*, 1982, **45**, 689.
196. THENO, D. M., SIEGEL, D. G. and SCHMIDT, G. R. *J. Food Sci.*, 1978, **43**, 483.
197. THENO, D. M., SIEGEL, D. G. and SCHMIDT, G. R. *J. Food Sci.*, 1978, **43**, 493.
198. THOMAS, D. J. and WAGNER, M. K. *J. Food Sci.*, 1987, **52**, 793.
199. THOMSON, J. E., BAILEY, J. S. and COX, N. A. *Poultry Sci.*, 1979, **58**, 139.
200. TIMS, M. and WATTS, B. M. *Food Technol.*, 1958, **12**, 240.
201. TOMPKIN, R. B. *J. Food Safety*, 1984, **6**, 13.
202. TOTH, L. and HAMM, R. *Fleischwirtschaft*, 1968, **48**, 1651.
203. TROUT, G. R. and SCHMIDT, G. R. *Proc. Rec. Meat Conf.*, Vol. 36, National Live Stock and Meat Board, Chicago, 1983, p. 24.
204. TROUT, G. R. and SCHMIDT, G. R. *J. Food Sci.*, 1984, **49**, 687.
205. TROUT, G. R. and SCHMIDT, G. R. *J. Food Sci.*, 1986, **51**, 1416.
206. TROUT, G. R. and SCHMIDT, G. R. *J. Agric. Food Chem.*, 1986, **34**, 41.
207. USDA. *Fed. Reg.—Rules and Reg.*, 1982, **47**, 10779.
208. VAAMONDE, G. and CHIRIFE, J. *Intl J. Food Microbiol.*, 1986, **3**, 51.
209. VAN WAZER, J. R. *Phosphates in Food Processing* (eds J. M. Deman and P. Melnychyn), AVI Publishing Co., Westport, 1971, p. 1.
210. VAN WAZER, J. R. and CAMPANELLA, D. A. *J. Amer. Chem. Soc.*, 1950, **72**, 655.
211. VAN WAZER, J. R. and HOLST, K. A. *J. Amer. Chem. Soc.*, 1950, **72**, 639.
212. VAN WAZER, J. R. and CALLIS, C. F. *Chem. Rev.*, 1958, **58**, 1011.
213. VENUGOPAL, V., PANSARE, A. C. and LEWIS, N. F. *J. Food Sci.*, 1984, **49**, 1078.
214. VOLLMAR, E. K. and MELTON, C. C. *J. Food Sci.*, 1981, **46**, 317.
215. WAGNER, M. K. *J. Food Prot.*, 1986, **49**, 482.
216. WAGNER, M. K. and BUSTA, F. F. *J. Food Sci.*, 1983, **48**, 990.
217. WAGNER, M. K. and BUSTA, F. F. *J. Food Sci.*, 1984, **49**, 1588.
218. WAGNER, M. K. and BUSTA, F. F. *Appl. Environ. Microbiol.*, 1985, **50**, 16.
219. WAGNER, M. K. and BUSTA, F. F. *J. Food Prot.*, 1985, **48**, 421.
220. WAGNER, M. K. and BUSTA, F. F. *J. Food Prot.*, 1985, **48**, 693.
221. WAGNER, M. K. and BUSTA, F. F. *J. Food Prot.*, 1986, **49**, 352.
222. WATSOS, E. Ph.D. thesis, 1981, Ludwig-Maximilians-University, Munich.
223. WATTS, B. M. *Adv. Food Res.*, 1954, **5**, 1.
224. WHITING, R. C. *J. Food Sci.*, 1984, **49**, 1355.

225. WHITING, R. C., BENEDICT, R. C., KUNSCH, C. A. and WOYCHICK, T. *J. Food Sci.*, 1984, **49**, 351.
226. WHITING, R. C., BENEDICT, R. C., KUNSCH, C. A. and BLALOCK, D. *J. Food Sci.*, 1985, **50**, 304.
227. WIERBICKI, E., KUNKLE, L. E. and DEATHERAGE, F. E. *Fleischwirtschaft*, 1963, **15**, 404.
228. WOYEWODA, A. D. and BLIGH, E. G. *J. Food Sci.*, 1986, **51**, 932.
229. YASUI, T., SAKANISHI, M. and HASHIMOTO, Y. B. *J. Agric. Food Chem.*, 1964, **12**, 392.
230. YASUI, T., FUKAZAWA, T., TAKAHASHI, K., SAKANISHI, M. and HASHIMOTO, Y. B. *J. Agric. Food Chem.*, 1964, **12**, 399.
231. YOUNG, L. L., LYON, C. E., SEARCY, G. K. and WILSON, R. L. *J. Food Sci.*, 1987, **52**, 571.
232. ZAKULA, R. *Proc. Europ. Meet. Meat Res. Work.*, 1969, **15**, 157.
233. ZEMEL, M. B. *J. Food Sci.*, 1984, **49**, 1562.
234. ZEMEL, M. B. and BIDARI, M. T. *J. Food Sci.*, 1983, **48**, 567.
235. ZEMEL, M. B., DAVIS, D. J. and PROULX, R. S. *Fed. Proc.*, 1984, **43**, 1052.
236. ZEMEL, M. B. and LINKSWILER, H. M. *J. Nutr.*, 1981, **111**, 315.
237. ZEMEL, M. B., SOULLIER, B. A. and STEINHARDT, N. J. *Fed. Proc.*, 1983, **42**, 397.

Chapter 7

HEAT TRANSFER DURING FREEZING AND THAWING OF FOODS

J. Succar

*Research and Development Center, Beatrice/Hunt-Wesson, Inc.,
California, USA*

SUMMARY

*Fundamental laws governing the phenomena of heat transfer during freezing
and thawing including the types and treatment of boundary conditions and
thermophysical properties are critically discussed. Empirical equations known
to predict reliably the functional relationships of thermal conductivity,
density, effective specific heat, and enthalpy with freezing or defrosting
temperatures, are proposed for their use in the development of versatile finite
difference techniques to predict transient state temperatures during commercial
operations.*

*Discussions are presented on the status of available procedures to predict
freezing and thawing times, and the need and importance of conducting
parametric analysis to further understand the phenomenon of heat transfer
with phase change.*

*Finally, two finite difference approaches to predict non-symmetric heat
transfer (including radiative heat exchange) during freezing and thawing are
critically presented. These techniques should be of interest to engineers
involved in the design and optimization of freezing or defrosting equipment and
processes.*

NOTATION

Bi	Biot number $= hl/k$
c	Apparent specific heat of frozen food (kcal/kgC°)
c_e	Empirical constant in eqn (23) (kcal/kgC°)
c_1	Specific heat of thawed food (kcal/kgC°)
C_{eh}	Dimensionless group for predicting apparent specific heat of frozen food (Table 6)
D	Empirical constant in eqn (23) (kcal $\cdot$ C$^{°n-1}$/kg)
E	Convergency criterion for implicit finite differences (eqn (88))
F	Geometric shape factor for thermal radiation
Fo	Fourier time, eqn (36) (dimensionless)
h	Surface heat transfer coefficient (kcal/h $\cdot$ m$^2 \cdot$ C°)
H	Food enthalpy (kcal/kg)
H_c	Food enthalpy extrapolated to $T_r = -40$°C (kcal/kg)
k	Thermal conductivity (W/mC° or kcal/h $\cdot$ m $\cdot$ C°)
k_l	Thermal conductivity of unfrozen food (W/mC° or kcal/h $\cdot$ m $\cdot$ C°)
k_r	Reference thermal conductivity of frozen food determined empirically with eqn (24) (W/mC° or kcal/h $\cdot$ m $\cdot$ C°)
k_{lx}, k_{ly}	x-directional or y-directional thermal conductivity at T_{sh} (W/mC° or kcal/h $\cdot$ m $\cdot$ C°)
k_{rx}, k_{ry}	Reference x-directional or y-directional thermal conductivity of frozen food determined empirically with eqn (24) (W/mC° or kcal/h $\cdot$ m $\cdot$ C°)
l	Food thickness (m)
L	Latent heat (kcal/kg)
L_r	Latent heat of sublimation of surface moisture at T_{ar} (kcal/kg)
$M_d, M_{klx}, M_{kly}, M_{rx}$	Dimensionless groups defined in Table 6
m	Partial pressure of water in surrounding atmosphere (kPa)
m_r	Value of m at T_{ar} (kPa)
n	Empirical constant in eqn (23), for predicting apparent specific heat (dimensionless)
n_{wo}	Fractional water content of unfrozen food
N	Number of nodal points for finite differences, Fig. 10

p	$= 0$ for rectangular foods
	$= 1$ for cylindrical foods
	$= 2$ for spherical foods
q	Heat flux (kcal/h)
R_c, R_d, R_{rx}	Dimensionless groups defined in Table 6
S_d, S_k, S_l	Constants for estimating k or ρ of food with eqn (24), (25), or (28). Appropriate units should be used. (W/mC$^{\circ 2}$, or kg/m^{3}C$^\circ$)
S_f	Shape factor. For *rectangular* food $S_f = x$-directional length/y-directional length. For *cylindrical* food $S_f =$ height/radius
S_{lx}, S_{ly}	Rate of change of k for $T > T_{sh}$, in the x-direction and the y-direction, respectively. For axisymmetric heat conduction S_{lx} refers to S in the radial direction. Both terms are equivalent to S_l in eqn (28) (W/mC$^{\circ 2}$)
S_{sx}	Rate of change of k for $T \leqslant T_{sh}$ in the x-direction for rectangular foods, or radial directions for cylindrical foods. Equivalent to S_k in eqn (24) (W/mC$^{\circ 2}$)
t	Time (h)
T	Temperature ($^\circ$C)
T^A	Absolute temperature (K)
T_a	Ambient temperature, beyond thermal boundary layer ($^\circ$C)
T_{ar}	Final reference temperature. $T_{ar} = -40^\circ$C for freezing and $T_{ar} = +8^\circ$C for thawing
T_c or T_f	Food temperature at thermal center ($^\circ$C)
T_r	Reference food temperature for calculating enthalpy with eqn (29) ($T_r = -40^\circ$C)
T_{or}	Initial reference temperature. $T_{or} = +8^\circ$C for freezing and -40°C for thawing
U	Dimensionless temperature. $U = (T - T_{ar})/(T_{or} - T_{ar})$
U_o	Initial food temperature (dimensionless)
U_a	Environmental food temperature (dimensionless)
U_e	Food temperature of radiative heat sink (dimensionless)
U_f	Final food temperature at thermal center (dimensionless)
U_{sh}	Highest freezing temperature of food (dimensionless)

V_i	Defined with eqn (63) (kcal/m^{3-p}°C)
x, y, z	Coordinate axes
z_i	Defined with eqn (79) (m^{p+1})
α_i	Thermal diffusivity constant for finite differences, defined with eqn (50) or (91) (m^2/h)
β_i, γ_i	In explicit finite differences these terms refer to dimensionless constants related to inverse of Biot numbers, defined with eqn (52) and (51), respectively
γ	In Table 5 or 6, $\gamma = k_{ry}/k_{rx}$
γ_i	In implicit finite differences defined with eqn (65) (W/m^{1-p}C°)
ε	Effective emissivity for thermal radiation ($0 \le \varepsilon \le 1$)
η	In finite differences defined with eqn (64) (W/m^{1-p}C°)
θ	Weight factor for implicit finite differences, eqn (62)
λ	In finite differences defined with eqn (80) (W/m^{1-p}C°)
μ	In finite differences defined with eqn (74) (dimensionless)
ρ	Food density (kg/m^3)
ρ_1	Density of unfrozen food (kg/m^3)
ρ_r	Empirical constant in eqn (25) (kg/m^3)
σ	Stephan-Boltzmann constant (W/m^{2}K^4)
ψ	In implicit finite differences defined with eqn (75) (dimensionless)
ω	In implicit finite differences defined with eqn (76) (h/m^2)
Δt	Time increment for finite differences (h)
Δx	Space increment for finite differences, defined by eqn (45) (m)
$]_s$	Evaluated at condition given by s

Subscripts

a	Ambient condition
abs	Absolute (i.e. $T_{abs} = 273 \cdot 16$°K)
d	Density
e	Heat sink for thermal radiation
eo, el	Radiative heat transfer to the left and right hand sides of infinite slabs, or center and exposed surface of cylinders and spheres (Fig. 3 and Fig. 11)
f	Final freezing or thawing temperature evaluated at the thermal center
$i \pm b$	For $b = 0$, $\frac{1}{2}$, or 1; refers to value of thermophysical property evaluated at that i location during finite differences
k	Thermal conductivity
l	Right hand side surface of infinite slabs, or exposed surface of cylinders and spheres

mn, mu	Mass transfer properties of two-dimensional rectangular and cylindrical foods (Fig. 9)
n	Same as subscript qn
o	Left hand side of infinite slabs or center infinite cylinders and spheres
q	Convective heat transfer
qn, qu	Heat transfer properties of two-dimensional rectangular and cylindrical foods (Fig. 9)
r	Radiative heat transfer
rb	Radiative heat transfer of black body
s	Food surface
sh	Highest freezing point
so, sl	Surfaces to the left and right hand side of infinite slabs, or center and exposed surfaces of infinite cylinders and spheres
sw	Freezing point of water
u	Same as subscript qu
x, y, z	Coordinate axes

Superscript

A	Absolute temperature
t	Refers to current time step in finite differences
$t+1$	Refers to advanced time step in finite differences $(t+1=t+\Delta t)$
$k, k+1$	Order of the Jacobi iteration procedure during estimation of $[\vec{T}^{t+1}]$ with eqn (83) (see eqn (87))
$\rightarrow$	Vector
$\rightrightarrows$	Matrix

1. INTRODUCTION

Understanding the phenomena of heat transfer in foods undergoing phase change continues to be the focus of attention of food engineers in the academic and industrial communities. The ultimate goal appears to be the development of simplified equations to predict freezing or thawing times. Accurate predictions are necessary for a variety of reasons, including designing and operating freezers, ovens, and storage rooms, or calculating freezing rates, and product throughputs. The phenomena of heat transfer during phase change is by no means simple. During melting or solidification the thermophysical properties of foods undergo sharp changes which follow non-linear relationships with freezing temperatures. Simplified solutions will generally require simplified assumptions on thermophysical properties, boundary conditions and operational conditions. Most of the simplified equations are improvements of Plank's equation[1] published in 1941. Even though Plank imposed restrictive assumptions on food properties and operational conditions, his model is widely used because of its mathematical simplicity. It is applicable only to estimate freezing or defrosting time and does not provide information on transient temperature distributions. The mathematical models based on the improvement of Plank's equation have introduced numerical modifications to account for the restrictive assumptions considered by Plank; therefore, these models are more reliable. Other methods to predict freezing and thawing times are based on solutions of Fourier's heat conduction equation. Approximate analytical solutions with less restrictive assumptions have been derived through application of Goodman's[2] heat balance integral technique.

Attempts to develop simple formulae to predict freezing and thawing times are numerous. Cleland and Earle[3] noted that in spite of the numerous methods proposed, there is no definite predictive model which can be used with confidence by food engineers. Situations of conflict arise when a given method performs well in certain freezing or thawing conditions but not in others.

The most reliable approach for predicting freezing or thawing times is the use of numerical solutions. The most widely used numerical techniques are finite differences and finite elements. These techniques can be used to solve the Fourier heat conduction equation with realistic assumptions on the thermophysical properties of the food, boundary conditions, operating conditions, and product shapes.

2. THE HEAT TRANSFER EQUATION APPLIED TO FREEZING AND DEFROSTING

2.1. Deriving the Governing Partial Differential Equation

During freezing, ice formation begins when the food temperature reaches the point of initial crystallization, also called the highest freezing temperature, T_{sh}. This initial freezing requires the release of a large amount of energy in the form of latent heat of phase change, to form the ice crystals. The remaining water which does not freeze at T_{sh} contains a higher concentration of food solubles and will freeze subsequently as the temperature continues to decrease.[4-10] This crystallization/concentration phenomenon will continue with decreasing temperatures. Since the water fraction in the remaining unfrozen food is smaller, the latent heat for subsequent crystallization will decrease, until a eutectic temperature is reached, below which no further crystallization occurs.

Researchers have found that if the latent heat of phase change is incorporated into the temperature-dependent thermophysical property effective (or apparent) specific heat c, heat transfer during freezing and thawing may be treated as a regular transient heat conduction problem. Therefore, when the effective specific heat includes the latent heat of crystallization or solidification, heat transfer with phase change may be predicted with the well-known Fourier heat conduction equation. Derivation of this equation is simple if we consider an infinitesimal volumetric element of food as shown in Fig. 1. In this element heat is entering the food through three planes at the rates q_x, q_y and q_z, moving in a direction parallel to the coordinate axis x, y and z, respectively.

Similarly, we assume that heat is leaving the element at the rates q_{x+dx}, q_{y+dy} and q_{z+dz}. Therefore, the heat accumulated in the volumetric element is given by:

$$\text{Heat in} - \text{heat out} = \text{Heat accumulated}$$

$$(q_x dy\, dz - q_{x+dx} dy\, dz) +$$

$$(q_y dx\, dz - q_{y+dy} dx\, dz) + {} = \rho c \frac{\partial T}{\partial t} dx\, dy\, dz \tag{1}$$

$$(q_z dx\, dy - q_{z+dz} dx\, dy)$$

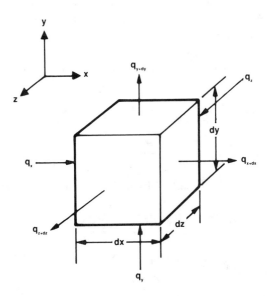

FIG. 1. Volumetric element of food showing three-dimensional heat transfer.

The terms ρ and c in eqn (1) are the thermal dependent density and apparent specific heat of the food. In cgs units the apparent specific heat is the amount of heat (calories) needed to increase the temperature of one gram of mass by 1°C. As indicated earlier, for freezing and thawing c also includes the heat needed for phase change.

Equation (1) may be rearranged as follows:

$$(q_x - q_{x+dx})\, dy\, dz + (q_y - q_{y+dy})\, dx\, dz +$$
$$(q_z - q_{z+dz})\, dx\, dy = \rho c \frac{\partial T}{\partial t} dx\, dy\, dz \tag{2}$$

The terms q_{x+dx}, q_{y+dy} and q_{z+dz} in eqn (2) are unknown functions of x, y and z, respectively. These terms may be approximated by Taylor's series.

$$q_{x+dx} = q_x + \frac{\partial q_x}{\partial x} dx + \frac{\partial^2 q_x}{\partial x^2} \frac{(dx)^2}{2!} + \frac{\partial^3 q_x}{\partial x^3} \frac{(dx)^3}{3!} + \cdots \tag{3}$$

$$q_{y+dy} = q_y + \frac{\partial q_y}{\partial y} dy + \frac{\partial^2 q_y}{\partial y^2} \frac{(dy)^2}{2!} + \frac{\partial^3 q_y}{\partial y^3} \frac{(dy)^3}{3!} + \cdots \tag{4}$$

$$q_{z+dz} = q_z + \frac{\partial q_z}{\partial z} dz + \frac{\partial^2 q_z}{\partial z^2} \frac{(dz)^2}{2!} + \frac{\partial^3 q_z}{\partial z^3} \frac{(dz)^3}{3!} + \cdots \tag{5}$$

Since dx, dy and dz tend to zero, Taylor's series converge rapidly, and only the first two terms in the right hand side of eqns (3), (4) and (5) suffice to describe q_{x+dx}, q_{y+dy} and q_{z+dz}.

$$q_{x+dx} = q_x + \frac{\partial q_x}{\partial x} dx \rightarrow (q_x - q_{x+dx}) = - \frac{\partial q_x}{\partial x} dx \tag{6}$$

$$q_{y+dy} = q_y + \frac{\partial q_y}{\partial y} dy \rightarrow (q_y - q_{y+dy}) = - \frac{\partial q_y}{\partial y} dy \tag{7}$$

$$q_{z+dz} = q_z + \frac{\partial q_z}{\partial z} dz \rightarrow (q_z - q_{z+dz}) = - \frac{\partial q_z}{\partial z} dz \tag{8}$$

Multiplying eqns (6), (7) and (8) by $dydz$, $dxdz$ and $dxdy$, respectively:

$$(q_x - q_{x+dx})\, dy\, dz = - \frac{\partial q_x}{\partial x} dx\, dy\, dz \tag{9}$$

$$(q_y - q_{y+dy})\, dx\, dz = - \frac{\partial q_y}{\partial y} dx\, dy\, dz \tag{10}$$

$$(q_z - q_{z+dz})\, dx\, dy = - \frac{\partial q_z}{\partial z} dx\, dy\, dz \tag{11}$$

Replacing eqns (9), (10) and (11) in eqn (2):

$$\rho c \frac{\partial T}{\partial t} = -\left(\frac{\partial q_x}{\partial x} + \frac{\partial q_y}{\partial y} + \frac{\partial q_z}{\partial z}\right) \tag{12}$$

Finally, substituting the heat flux terms q_x, q_y and q_z in eqn (12) by the Fourier of heat conduction ($q = -k\partial T/\partial x$), the three-dimensional form of the Fourier heat conduction equation is obtained.

$$\rho c \frac{\partial T}{\partial t} = \frac{\partial}{\partial x}\left(k\frac{\partial T}{\partial x}\right) + \frac{\partial}{\partial y}\left(k\frac{\partial T}{\partial y}\right) + \frac{\partial}{\partial z}\left(k\frac{\partial T}{\partial z}\right) \tag{13}$$

Equation (13) applies to any point of the solid assuming that no heat is generated at that point. Furthermore, the solid does not have to be homogeneous or isotropic.

Derivations similar to the ones used to obtain eqn (13) may be found in textbooks by Carslaw and Jaeger,[11] Luikov,[12] Holman,[13] Fahien[14] and others. Even though our derivation is for three-dimensional cartesian coordinate variables, the geometrical shapes of a cylinder, ellipsoid, and sphere may also be easily defined by using cylindrical, ellipsoidal and spherical coordinates.[15] The form of eqn (13) is appropriate for predicting heat transfer during phase change, since it allows for the use of the highly non-linear temperature dependence of the density apparent specific heat and thermal conductivity of the food.

2.2. Proposing the Initial and Boundary Conditions

Before solving the heat conduction equation it is necessary to define the initial and boundary conditions. Initial conditions describe the temperature of the food at the beginning of the process. Boundary conditions supply information on external conditions surrounding the food, affecting the behavior of the surface food temperature and heat transfer rates during freezing or thawing.[15]

2.2.1. Initial Conditions

Initial conditions usually assume that the temperature of the food is uniform throughout its mass. In general, the initial condition for the temperature distribution can be defined as a location dependent function f at the beginning of the process (time $t=0$):

$$T = f(x, y, z) \quad \text{at } t = 0 \tag{14}$$

2.2.2. Boundary Conditions

Boundary conditions of interest to food engineers are four in number.

2.2.2.1. Prescribed surface temperature. In this condition the surface temperature is identical to that of the heating or cooling medium, and remains constant throughout the process. Bakal and Hayakawa[16] indicate that this condition may arise during freezing in plate freezers, or in well stirred liquids. For example, this boundary condition applied to the right hand side of the infinite slab shown in Fig. 2 would be:

$$T_s = T_a \text{ at } x = l \text{ for } t > 0 \tag{15}$$

Where T_s and T_a are the food surface and surrounding medium temperatures, respectively.

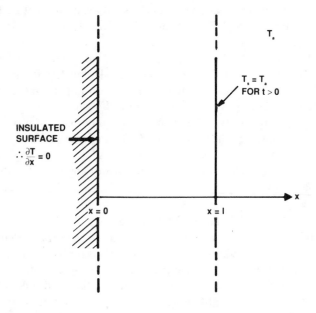

FIG. 2. Schematic representation of an infinite slab, showing boundary conditions 1 and 2.

2.2.2.2. No flux across the surface. This condition describes an insulated surface, or the center of a slab heated or cooled symmetrically:

$$\frac{\partial T}{\partial x} = 0 \quad \text{at } x = 0 \tag{16}$$

Differentiation is in a direction outward and normal to the surface of interest.[11] Figure 2 illustrates the case of insulating one surface of an infinite slab.

2.2.2.3. Linear heat transfer at the surface. This condition assumes that the surface temperature is different from the temperature of the surrounding medium T_a, and that heat fluxes are proportional to the varying temperature difference between this surface and T_a.

$$k \frac{\partial T_s}{\partial x} = -h(T_s - T_a) \qquad (17)$$

The proportionality constant h is referred to as the surface conductance or the coefficient of surface heat transfer. The surrounding medium temperature T_a must be measured at or beyond the thermal boundary layer. This temperature is also called the bulk temperature (Fig. 3).

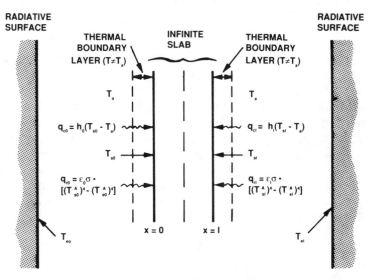

FIG. 3. Non-symmetric convective and radiative boundary conditions imposed on the surfaces of an infinite slab.

2.2.2.4. Radiation heat transfer. This condition indicates that heat is transferred through electromagnetic radiation, as a result of differences between the temperature on the food surface and the temperature on the surfaces of the surrounding heating or cooling medium. This phenomenon is called thermal radiation, and implies that heat may also be transferred through regions where a perfect vacuum exists (i.e. $h = 0$). In the electromagnetic spectrum, thermal radiation lies in the range from about 0·1 to 100 micrometers, and includes a portion of infrared light, all visible light, and a portion of ultraviolet light. The ideal radiator or black body emits energy at a rate proportional to the fourth power of its absolute temperature, T^A:

$$q_{rb} = \sigma(T^A)^4 \tag{18}$$

Equation (18) is the Stefan–Boltzmann law. The term q_{rb} is the energy radiated by an ideal radiator (W/m^2), and σ is the Stefan–Boltzmann constant which has the value:

$$\sigma = 5·669 \times 10^{-8} \text{ W/m}^2 \text{ K}^4 \qquad 0·1714 \times 10^{-8} \text{ Btu/h ft}^2 \text{ R}^4$$

Radiation heat exchange between two bodies with different temperatures is also proportional to the fourth power of their absolute surface temperatures:

$$q_{rb} = F\sigma[(T_1^A)^4 - (T_2^A)^4] \tag{19}$$

The term F in eqn (19) is the geometric shape factor for thermal radiation. It is a measure of the fraction of energy leaving the surfaces of the heating or cooling medium which reaches the surfaces of the food. Since foods are enclosed within the heating or cooling medium, the fractional value of F should approach unity.[17,18]

Equation (19) is only applicable to radiative heat exchange between ideal black bodies, which are assumed to absorb all incident radiation (absorptivity, $\alpha = 1·0$); no radiation is reflected, none is transmitted. Foods, however, behave as 'gray' bodies,[17,21] in which $0 < \alpha < 1$. To account for the 'gray' nature of foods, it is necessary to introduce an effective emissivity term ε. In these cases:

$$q = F\varepsilon\sigma[(T_1^A)^4 - (T_2^A)^4] \tag{20}$$

During freezing and thawing foods are usually exposed to boundary conditions 3 and 4. Figure 3 illustrates an infinite slab undergoing non-symmetric heat exchange. Notice that the surface heat transfer coefficients and the effective emissivities are different on each face of the slab. The boundary conditions may then be mathematically expressed as shown below (refer to Fig. 3).

Left side of slab:

$$k\frac{\partial T_{so}}{\partial x}\bigg]_{x=0} = -h_o(T_{so} - T_a) - \varepsilon_o\sigma[(T_{so}^A)^4 - (T_{es}^A)^4] \tag{21}$$

Right side of slab:

$$-k\left.\frac{\partial T_{sl}}{\partial x}\right]_{x=l} = -h_l(T_{sl}-T_a)-\varepsilon_l\sigma[(T_{sl}^A)^4-(T_{el}^A)^4] \qquad (22)$$

Radiation heat transfer is seldom considered for the solution of heat transfer with phase change. Instead, researchers determine a bulk surface heat transfer coefficient which includes convection and radiation. However, Succar and Hayakawa[19,20] have reported that under certain conditions radiative heat exchange, eqn (20), is a significant fraction of the total heat entering or leaving the food. Since radiative heat exchange is a highly non-linear function of temperature, a single lumped term will not properly predict the transient state heat conduction in foods. For this reason, Succar and Hayakawa[20] developed a Response Surface Method for the simultaneous estimation of convective heat transfer coefficients and effective emissivities during freezing and thawing of foods.

3. THERMOPHYSICAL PROPERTIES DURING FREEZING AND THAWING

Reliable solutions to the heat conduction equations with phase change can only be obtained when realistic boundary conditions and thermophysical properties are used. Thermophysical properties needed for these solutions are the apparent specific heat c, the thermal conductivity k, and the density ρ. As discussed below, for these predictions one needs two additional properties: the functional relationship of the food's enthalpy H versus temperature, and the highest freezing temperature T_{sh}, also known as the temperature of initial crystallization. Figures 4–7 describe the functional relationship of H, c, k and ρ with temperature. Inspection of these figures show that H, c, k and ρ are highly non-linear below T_{sh}. Above T_{sh} the food is thawed, and it may be assumed that only k and H change linearly with temperature, while c and ρ remain virtually constant.

Theoretical and empirical methods have been developed to predict the thermophysical properties of the food in the freezing range.[22] The theoretical equations usually consider the thermophysical properties of the food components, and/or assume that, thermodynamically, the food behaves as an ideal binary solution system. Empirical equations on the other hand are developed to fit thermophysical data obtained through experimental procedures, using mathematical and statistical techniques.

Theoretical equations to predict thermophysical properties have been reported by Chen,[7-10] Heldman,[23,24] Heldman and Gorby,[25] Hohner and

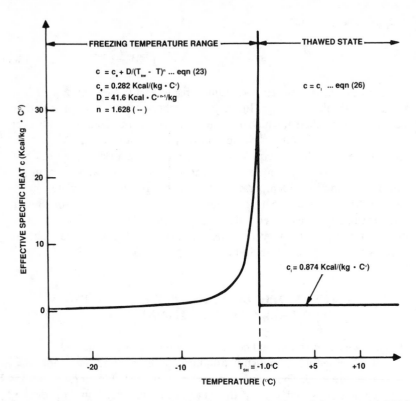

FIG. 4. Functional relationship between c and T for lean fish meat with 75% moisture. Estimated with Riedel's enthalpy data[42] (Fig. 4) and modified Schwartzberg formula eqn (23) or eqn (26).

Heldman,[26] Hsieh et al.,[27] Lescano and Heldman,[28] Schwartzberg,[29,30] Schwartzberg et al.,[31] Levy[32-34] and Jason and Long.[35] These methods are important since they can be used to approximate the behavior of the thermophysical properties of the frozen food when experimental data are not available. However, Succar and Hayakawa[36] caution that there may be significant differences between predicted and experimental thermophysical properties when one used theoretical methods based on simplified assumptions on the thermodynamic behavior of foods. Instead, empirical equations known to predict reliably the thermophysical properties of a food system should be used. The availability of empirical equations in the literature is scarce. Empirical correlations to predict the enthalpy of food systems were proposed by Riedel[37] and Chang and Tao.[38] Ramaswany and Tung[39]

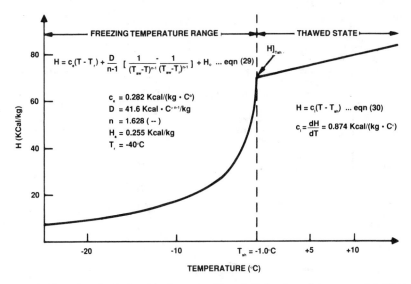

FIG. 5. Functional relationship between H and T, for lean fish meat with 75% moisture.[42]

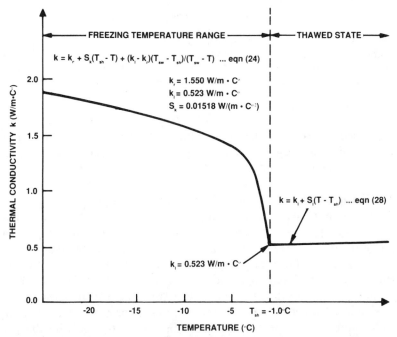

FIG. 6. Thermal conductivity of cod and haddock[35] in the freezing temperature range, as predicted by Schwartzberg's eqn (24). Empirical parameters k_r and S_k obtained through a non-linear parametric analysis.[36]

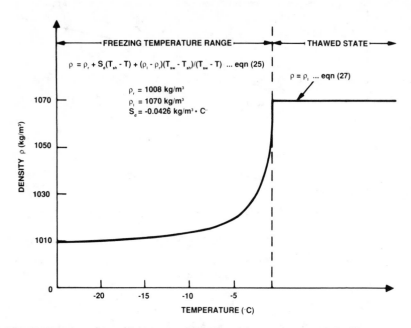

FIG. 7. Density of lean fish meat with 75% moisture. To obtain the functional relationship between ρ and T, semi-experimental values were obtained with eqn (31) using f_i values from Riedel;[42] these values were then fitted with eqn (25) using empirical constants ρ_r and S_d obtained through a non-linear parameter estimation technique.[36]

developed empirical correlations to estimate the thermophysical properties of apples during freezing. Schwartzberg[29] developed empirical equations to predict the thermal conductivity of foods in the freezing range.

3.1. Equations for Predicting Thermophysical Properties

Because of the need to predict accurately the values of c, k and ρ, Succar and Hayakawa[36] proposed empirical equations to predict these thermophysical properties during freezing and defrosting temperatures. The equations for enthalpy apparent specific heat and thermal conductivity in the freezing temperature range, were slight modifications of Schwartzberg's theoretical formulae,[29,30] while the equation for density was mathematically designed to simulate the behavior of this thermophysical property in the freezing temperature range. These equations were shown to be highly reliable and are presented below.

For $T \leqslant T_{sh}$:

$$c = c_e + D/(T_{sw} - T)^n \tag{23}$$

$$k = k_r + S_k(T_{sh} - T) + (k_1 - k_r)\left[\frac{T_{sw} - T_{sh}}{T_{sw} - T}\right] \tag{24}$$

$$\rho = \rho_r + S_d(T_{sh} - T) + (\rho_1 - \rho_r)\left[\frac{T_{sw} - T_{sh}}{T_{sw} - T}\right] \tag{25}$$

For $T > T_{sh}$:

$$c = c_1 \tag{26}$$

$$\rho = \rho_1 \tag{27}$$

$$k = k_1 + S_1(T - T_{sh}) \tag{28}$$

Equations (23), (24) and (25) needed to predict c, k and ρ for $T \leqslant T_{sh}$ require the following empirical constants:

Apparent specific heat c: c_e, D and n
Thermal conductvity k: k_r and S_k
Density ρ: ρ_r and S_d

Thermophysical values in these equations are:

T_{sh} = Highest freezing temperature
k_1 = Thermal conductivity at $T = T_{sh}$
ρ_1 = Density of the thawed food ($T > T_{sh}$)

For $T > T_{sh}$ the only empirical constant needed is S_1 in eqn (28), which may be readily obtained through linear regression analysis on thermal conductivity data of thawed food. To estimate the empirical constants for eqns (23), (24) and (25) Succar and Hayakawa[36] used a computerized non-linear parameter estimation procedure developed by Metzler et al.[40] Estimating the empirical constants for apparent specific heat, c_e, D and n requires special treatment. Equations (23) and (26) must first be integrated to develop expressions for enthalpy. For $T \leq T_{sh}$ the enthalpy equation will contain the three empirical constants needed for eqn (23). This approach for determining c_e, D and n is required since experimental published data for enthalpy are more reliable and available than data for apparent specific heat. Integrating eqns (23) and (26) delivers:

For $T \leq T_{sh}$:

$$H = \int_{T_r}^{T} c\, dT = c_e(T - T_r) + \frac{D}{(n-1)}\left[\frac{1}{(T_{sw} - T)^{n-1}} - \frac{1}{(T_{sw} - T_r)^{n-1}}\right] + H_c \qquad (29)$$

For $T > T_{sh}$:

$$H = \int_{T_{sh}}^{T} c\, dT = c_1(T - T_{sh}) + H]_{T_{sh}} \qquad (30)$$

where $H]_{T_{sh}}$ indicates the value of H evaluated at T_{sh}.

Equation (29) is valid to predict enthalpy values when $T \leq T_{sh}$. This equation contains the parameters of interest c_e, D and n, in addition to the parameter H_c. The term T_r is the reference temperature for calorimetric determinations of the enthalpy of foods, $T_r = -40°C$. Therefore, for $T_r = -40$ one should expect $H_c = 0$. However, Succar and Hayakawa[36] used eqn (29) with $H_c = 0$ to fit a variety of published H–T data determined experimentally. The results showed that when $H_c = 0$ the enthalpy values of fruits and sucrose solutions were underestimated by the same magnitude throughout the temperature range of the data. When a non-zero H_c was introduced in the analysis the correlation coefficients between experimental and predicted enthalpies were ≥ 0.998. Schwartzberg[41] contends that non-zero H_c values may be due to the fact that at critically low temperatures, the eutectic temperature for the solutes and water is reached, and eqn (29) no longer applies. For example, the eutectic temperature for sucrose and water is $-14.5°C$. If equilibrium is attained the system completely solidifies, resulting in an abrupt change in enthalpy unaccounted for with eqn (29). The H_c term then becomes a fully empirical constant (see illustration in Fig. 8) which is obtained by extrapolating the enthalpy function of eqn (29) to $T_r = -40°C$. Therefore, when providing the empirical constants for eqn (29), the range of temperatures for which these constants apply must also be given. Table 1 shows the values for empirical constants c_e, D, n and H_c, as well as c_1 values for foods at different water contents (n_{wo}). The c_e, D, n and H_c values were obtained by applying non-linear regression analysis[40] to the bivariate data (enthalpy, temperature) obtained from Riedel[42-45] and Muller.[46] The parameter c_1 was obtained through linear regression since H is linearly related to temperature when $T > T_{sh}$. The temperature of initial crystallization T_{sh} needed for the regression analyses was determined at the intersect of the non-linear H function in the freezing

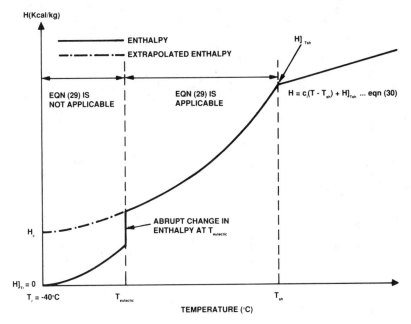

FIG. 8. Hypothetical enthalpy–temperature curve, showing a eutectic point at $T_{eutectic} > T_r$.

temperature range, and the linear H function in the thawed temperature range, following a mathematical procedure developed by Succar and Hayakawa.[19,47]

Empirical constants k_r and S_k to predict thermal conductivities in the freezing temperature range with eqn (24) are given in Table 2. These constants were obtained by analyzing published thermal conductivity values by Hsieh et al.,[27] Jason and Long,[35] Lentz[48] and ASHRAE.[49]

Equation (24) used with the empirical constants in Table 2 will describe functional relationships similar to that shown in Fig. 6, i.e. the thermal conductivity will increase with decreasing temperatures below T_{sh}. This behavior is typical of most foods. The increase in k is in large part due to the higher thermal conductivity of the ice which forms in the food as the temperature decreases.[48] However, Sastry and Datta[50] have reported that the effective thermal conductivity of some porous frozen foods such as ice creams show a decrease in thermal conductivity with decreasing temperature below T_{sh}. It is likely that Schwartzberg's eqn (24) should be capable of

TABLE 1
EMPIRICAL PARAMETERS FOR THE PREDICTION OF ENTHALPY AND APPARENT SPECIFIC HEAT OF FOODS AT DIFFERENT MOISTURE CONTENTS

Product	n_{wo}^a (—)	T_{sh} (°C)	c_i (kcal/kg-C°)	H_c (kcal/kg)	D (kcal-C°$^{n-1}$)/kg	c_e (kcal/kg-C°)	n (—)	Correlation coefficient
						Parameters in eqn (23) or eqn (29)		
Tylose MH 1000	0·77	−0·60	0·874	0·215	29·2	0·339	1·696	1·000
Tylose MH 1000 + 1% NaCl	0·77	−0·95	0·880	0·231	38·7	0·266	1·597	1·000
$(-40°C \leq T \leq T_{sh})$								
Beef meat (lean)	0·74	−0·99	0·835	0·000	51·6	0·448	1·968	1·000
$(-40°C \leq T \leq T_{sh})$	0·70	−1·01	0·811	0·000	40·5	0·369	1·717	0·999
	0·63	−1·76	0·774	0·000	74·4	0·415	1·981	1·000
	0·57	−2·02	0·740	0·022	38·6	0·235	1·497	0·998
	0·45	−4·09	0·700	0·057	39·5	0·136	1·416	1·000
Fish meat (lean)	0·82	−0·80	0·916	0·000	49·3	0·448	1·999	1·000
$(-40°C \leq T \leq T_{sh})$	0·75	−1·00	0·874	0·255	41·6	0·282	1·628	0·999
	0·66	−1·95	0·817	0·000	83·5	0·388	1·934	1·000
	0·57	−2·96	0·777	0·000	98·9	0·374	1·933	1·000
Fruit and vegetable juices[b]	0·96	−0·39	0·968	7·117	30·8	0·366	1·928	0·999
$(-30°C \leq T \leq T_{sh})$	0·87	−1·38	0·933	14·772	85·2	0·415	1·896	1·000
	0·75	−3·19	0·863	24·366	117·1	0·465	1·750	1·000
	0·61	−6·98	0·772	38·179	165·2	0·328	1·675	1·000
Sucrose solutions	0·96	−0·21	0·978	6·745	16·9	0·428	1·903	1·000
$(-14°C \leq T \leq T_{sh})$	0·61	−4·36	0·781	46·257	140·2	0·145	1·799	1·000

[a] n_{wo} = Moisture fraction in food.
[b] Also applicable to fresh fruit or vegetables.

TABLE 2

EMPIRICAL PARAMETERS FOR THE PREDICTION OF THERMAL CONDUCTIVITY OF SOME FOODS AND TYLOSE WITH THE USE OF EQN (24)

Product	$n_{wo}{}^a$ (—)	T_{sh} (°C)	Parameters in eqn (24)		
			k_1 (W/m-$C°$)	k_r (W/m-$C°$)	S_k (W/m-$C°^2$)
Tylose MH1000^b	0·77	−0·6	0·490	1·578	0·007 73
Lean beef meat (‖)^c	0·75	−0·99	0·477	1·403	0·007 65
Lean beef meat (⊥)^c	0·74	−0·99	0·477	1·078	0·006 23
Codfish	0·82	−0·8	0·523	1·302	0·012 06
Cod and haddock	0·75	−1·0	0·523	1·550	0·015 18
Asparagus	0·93	−0·7	0·530	1·299	0·016 71
Strawberries	0·89	−0·9	0·540	1·950	0·014 60
Carrots	0·88	−1·1	0·500	1·284	0·022 14
Cherries	0·87	−1·4	0·530	1·718	0·019 39
Peas	0·76	−1·8	0·470	1·983	0·011 72
Plums	0·76	−2·3	0·510	2·316	0·001 47

$^a n_{wo}$ = Moisture fraction in food.
^{b} Karlsruhe Test Material.
^{c} ‖ signifies heat transferring in a direction parallel to muscle fibers, ⊥ in a direction perpendicular to muscle fibers.

predicting this irregular thermal conductivity behavior through using appropriate k_r and S_k terms determined through non-linear regression with experimental data.

There is a great scarcity of data on the densities of food undergoing freezing or defrosting in published articles. Some values have been reported by Hsieh et al.,[27] Ramaswamy and Tung,[39] and Dickerson.[51] Because of the limited density data available, Succar and Hayakawa[36] generated semi-experimental density values by using the following formula proposed by Heldman[24] and Hsieh et al.,[27] together with published Mollier charts (Riedel[42–45] and Muller[46]).

$$\frac{1}{\rho} = \frac{f_i}{\rho_i} + \frac{f_s}{\rho_s} + \frac{f_w}{\rho_w} \tag{31}$$

The terms f_i, f_s and f_w in eqn (31) are the fractions of ice, solids and water in the food at a given temperature; ρ_i, ρ_s and ρ_w are the densities of ice, solids and water respectively.

The density ρ_i, in eqn (31) is a function of freezing temperature, and was predicted with the following linear equation obtained through regression analysis on ice density data reported by Perry and Chilton:[52]

$$\rho_i = 916 \cdot 9 - 0 \cdot 156T \text{ for } -40°C \leqslant T \leqslant 0°C \tag{32}$$

In eqn (32), the temperature T is in °C and ρ_i is in kg/m^3. The varying fractions of ice and water in eqn (31) were obtained for fresh fruit, fresh vegetables, lean meat and lean fish meat, using Mollier charts. The effective densities of non-freezable solids, f_s, were estimated from published densities of the thawed product, since at these temperatures $f_i = 0$. The densities generated with eqn (31) were subjected to non-linear regression analysis,[40] to determine the value of the two empirical constants ρ_r and S_d in eqn (25). Table 3 shows typical values for the ρ_r and S_d applicable to tylose, meat, fish, fruit and vegetable juices, and sucrose solutions.

TABLE 3

EMPIRICAL PARAMETERS FOR THE PREDICTION DENSITY OF SOME FOODS AND TYLOSE

Product	n_{wo}[a] (—)	T_{sh} (°C)	Parameters in eqn (25)		
			ρ_l (kg/m³)	ρ_r (kg/m³)	S_d (kg/m³ − C°)
Tylose MH1000	0·77	−0·60	1 006	938	0·154 2
Lean beef meat	0·70	−1·01	1 070	1 012	0·005 6
	0·63	−1·76	1 075	1 019	0·202 7
	0·57	−2·96	1 080	1 025	0·332 2
	0·45	−4·09	1 090	1 057	0·110 4
Lean fish meat	0·82	−0·80	1 060	985	0·203 6
	0·75	−1·00	1 070	1 008	−0·042 6
	0·66	−1·95	1 075	1 016	0·200 0
	0·57	−2·96	1 080	1 030	0·243 7
Fruit and vegetable juices[b]	0·96	−0·39	880	817	0·134 8
	0·87	−1·38	943	943	0·107 9
	0·75	−3·19	1 100	1 030	−0·107 9
	0·61	−6·98	1 227	1 145	0·224 8
Sucrose solution	0·96	−0·21	1 000	921	0·056 6
	0·61	−4·36	1 009	957	0·273 5

[a] n_{wo} = Moisture fraction in food.
[b] Also applicable to fresh fruit or vegetables.

4. ALTERNATIVES FOR PREDICTING FREEZING AND THAWING TIMES OF FOODS

For the reliable prediction of transient state temperatures during phase change it is necessary to use realistic assumptions on thermophysical properties, boundary conditions and product shapes. However, when the thermophysical properties change with varying temperatures, no analytical solutions to the resultant non-linear heat conduction equations are available.[21,53-55] The solution is further complicated if radiative heat exchange is assumed at the boundaries. Due to the need to predict freezing and thawing times, researchers have proposed analytical solutions which were based on simplified assumptions on the shape of the food, thermophysical properties and boundary conditions. Some have assumed that phase change occurs at a single temperature, which departs from the known fact that foods freeze over a range of temperatures due to the presence of solutes. Others have developed predictive models using relationships obtained from heat transfer experiments and from analyses of theoretical heat balance equations.

An alternative way to predict freezing times has been the application of numerical finite difference or finite element methods. These techniques require the use of computers and are of special interest to the food engineer, since they can be used to obtain solutions to the heat transfer equations with phase change, for a variety of realistic assumptions on product shape, thermophysical properties and boundary conditions.

The mathematical methods developed for predicting freezing and thawing times are numerous. In 1977, Hayakawa[56] critically reviewed approximately 30 different methods. In 1985, this author and his coworkers[57] critically reviewed 25 additional mathematical methods. Since then, more contributions have been reported by Nonino and Hayakawa,[58] Wallhoever et al.,[59] Ilicali and Saglam,[60] Chan et al.,[61] Pham,[62,63] De Cindio et al.,[64] Taoukis et al.,[65] Lacroix and Castaigne.[66] Ramaswamy and Tung[67] cited 81 references in their critical review of published methods for predicting freezing times. An extensive list of 100 publications dealing with the prediction of heat transfer during phase change is reported by Holdsworth.[68]

4.1. Selecting A Predictive Model

In view of the large number of models available, the food engineer is faced with the challenging task of selecting a mathematical model to fit his/her own needs. This selection requires the careful examination of the thermophysical properties, boundary conditions and geometrical shapes assumed for the development of a given mathematical model.

Hayakawa *et al.*'s critical review[57] is very useful to select predictive models. In this review the available procedures are classified into empirical, semi-theoretical or theoretical methods. The empirical methods are based solely on empirical relationships derived from freezing or thawing experiments. The semi-theoretical methods are those based on the combined use of empirical relationships obtained from heat transfer experiments together with relationships derived through the analyses of theoretical heat balance equations. The theoretical methods are entirely based on mathematical solutions of theoretical heat transfer equations. In this review, the restrictive assumptions imposed for deriving the analytical or numerical solutions are given in a table form. These assumptions are classified into those concerning (1) boundary conditions, (2) initial conditions, (3) density of the sample object, (4) thermal conductivity and effective specific heat, and (5) convective and diffusional movement of unfrozen mass of solute due to its concentration gradients.

5. PARAMETRIC ANALYSES

Hayakawa *et al.*[17,18] and Succar and Hayakawa[69] noted that the operational conditions and thermophysical properties included in the development of the theoretical and semi-theoretical models were usually selected on the basis of experience and/or educated guesses as to which group of parameters had a significant influence in the rates of heat transfer during phase change. Up to 1983, except for the work of Hsieh *et al.*,[27] virtually no parametric analyses had been reported in the literature, in relation to the quantitative influence of thermophysical properties and operational conditions on the rates of heat transfer during freezing and thawing. Hayakawa proposed to use numerical finite differences and finite element solutions to the heat conduction equation with phase change, in conjunction with rigorous statistical procedures, to identify the parameters which have a significant influence in the phenomenon of heat transfer with phase change. His contributions are of great significance since with the numerical techniques Hayakawa and his coworkers were able to solve the equations for a variety of realistic assumptions on product shapes, thermophysical properties, and boundary conditions. Once the significant parameters influencing freezing or thawing were identified, they were used in the development of theoretical regression equations to predict freezing and thawing times over a wide range of parametric values found in industrial operations. The investigations resulting from these parametric analyses are listed in Table 4. Solutions to heat transfer equations in foods approximated by spheroids, finite cylinders, or rectangular shapes shown in Table 4, assumed that heat flux rates through the surfaces of the food were influenced by the rate of moisture evaporation or condensation. According to Holdsworth,[68] many products are frozen prior to packaging; in these cases the assumption that the rate of heat transfer is affected by mass transfer rates at the food boundaries is correct. Hayakawa's procedure for parametric analysis is elaborate but leads to reliable results. In this procedure the equations for heat conduction, boundary conditions, operational conditions (i.e. initial and final temperatures), and thermophysical properties, were transformed to dimensionless equations. With these transformations the number of independent parameters was reduced, thus simplifying computational analyses. From these new systems of equations all the dimensionless groups required to uniquely describe the freezing or thawing phenomena were identified through application of Buckingham pi theorem:[14,71] the number of dimensionless groups is equal to the number of variables minus the number of fundamental dimensions. The number of variables are the independent physical quantities (thermophysical properties, operational conditions, boundary conditions), and the fundamental dimensions are length, time, mass, temperature differential ($C°$), and temperature ($°C$). The number of dimensionless groups identified was usually large (i.e. 18 groups for infinite slabs under non-symmetric conditions, 15 groups for infinite cylinders and spheres, 22 groups for finite cylinders and rectangular shapes during freezing, and 23 groups for these two latter configurations during thawing). To further reduce the number of dimensionless variables, those which had a significant influence on freezing and thawing rates were identified using a statistical screening design[72,73] which enabled estimating the main effects of all parameters without confounding them with any interactions. With the design, a given number of freezing or thawing experiments were simulated for each food shape by using the proper combinations of minimal and/or maximal values of all parameters. The simulations were accomplished using proven computerized numerical procedures. The number of computer simulated experiments as well as parametric levels (minimal or maximal) needed for the design were obtained from Plackett and Burman,[72] and Kleijnen.[73] The minimal and maximals for each parameter were determined through extensive literature surveys on freezing and thawing conditions found during commercial operations. Major references for these surveys were Heldman and Singh,[5] Hsieh *et al.*,[27] Jason and Long,[35] Riedel,[42-45,74] Muller,[46] Lentz,[48] ASHRAE Handbook,[49] Dickerson,[51] and a set of reference books edited by

TABLE 4

PARAMETRIC ANALYSES ON FREEZING AND DEFROSTING BY HAYAKAWA AND COWORKERS

	Title—mathematical algorithm	Assumptions of model	Authors
(1)	Two-dimensional heat conduction in food undergoing freezing: development of computerized model. Used finite element method to solve the two-dimensional heat conduction equation	Anisotropic two-dimensional heat transfer in a parallelepiped or finite cylinder-shaped food. Convective and radiative non-symmetric boundary conditions, which included heat lost or gained due to moisture evaporation or condensation from exposed surfaces. Properties k and c are temperature dependent below and above T_{sh} (anisotropic k)	Hayakawa et al.[17]
(2)	Two-dimensional heat conduction in food undergoing freezing: predicting freezing time of rectangular or finitely cylindrical food. Used finite element method together with statistical techniques to develop predictive equations	Same as above	Hayakawa et al.[18]
(3)	Parametric analysis for predicting freezing time of infinitely slab-shaped food. Used fully implicit finite difference method together with statistical techniques to develop predictive equations	Food shape approximated by infinite slab. Convective and radiative heat transfer. Properties k, ρ and c are temperature dependent below and above T_{sh}	Succar and Hayakawa[69]

(4)	Parametric analysis of one-dimensional heat conduction in foods undergoing freezing and thawing. Same mathematical and statistical techniques as above	Food shape approximated by infinite slab, infinite cylinder or sphere. Convective and radiative heat transfer. Non-symmetric heat transfer for case of infinite slabs. Properties of k, ρ and c are temperature dependent below and above T_{sh}	Succar and Hayakawa[19]
(5)	Freezing or thawing time estimation of spheroidal food. Used finite element method together with statistical techniques to develop predictive equation	Food shape approximated by a spheroid which is a geometrical shape obtained by revolving an ellipse about one of its axes. Boundary conditions included convective and radiative heat exchange, and convective surface moisture exchange due to condensation, evaporation or sublimation. Properties k, ρ and c are temperature dependent below and above T_{sh}	Nonino and Hayakawa[70]
(6)	Thawing time of frozen food of a rectangular or finitely cylindrical shape	Anisotropic two-dimensional heat transfer in a parallelepiped- or finite cylinder-shaped food. Same assumptions as in Hayakawa et al.[17,18]	Nonino and Hayakawa[58]

Note: Thermophysical properties were predicted with eqns (23)–(30).

Tressler *et al.*[75] The dependent variables obtained with the screening designs were the freezing or thawing times. These dependent variables were subjected to analyses of variance[76] to identify significant and non-significant parameters. Fractional or full central composite designs[77-79] were applied to the significant parameters to develop algebraic regression formulae for predicting freezing or thawing times. For the development of these equations, the non-significant parameters were kept at a frequent value found during commercial operations, while the significant parameters assumed values within their maximals and minimals, according to the central composite designs selected.

For the parametric analyses of foods approximated by rectangular shapes or cylindrical shapes, the following parabolic heat transfer equation was proposed.[17,18]

$$x^p \rho c \frac{\partial T}{\partial t} = \frac{\partial}{\partial x}\left(x^p k_x \frac{\partial T}{\partial x}\right) + \frac{\partial}{\partial y}\left(x^p k_y \frac{\partial T}{\partial y}\right) \tag{33}$$

The value of $p = 0$ for axisymmetric heat conduction, otherwise $p = 1$.

Initial condition:

$$T = T_o \text{ when time } t = 0, \text{ for all } x \text{ and } y \tag{34}$$

Boundary condition (general form):

$$k_x e_x \frac{\partial T}{\partial x} + k_y e_y \frac{\partial T}{\partial y} = -h_q(T - T_a) - F\sigma\varepsilon[(T^A)^4 - (T_e^A)^4] - h_m L(m - m_r) \tag{35}$$

when $t > 0$ for all x and y on the boundary surface. The terms k_x and k_y in eqns (33) or (35) are the x- and y-directional thermal conductivities of food, needed to define anisotropic heat conduction; x is a radial directional variable when eqns (33)–(35) are applied to axisymmetric heat conduction (Fig. 9).

The first, second and third terms of the right hand side of eqn (35) signify convective surface heat transfer, radiative surface heat transfer, and surface moisture loss (or gain), respectively. The parameters in eqn (35) are:

e_x, e_y	Values of x- and y-directional cosines of an outward direction normal to the food surface boundary.
F	Geometric shape factor for thermal radiation; eqn (19). The value of F was assumed as 1.
h_q, h_m	Convective surface heat transfer coefficient and convective surface mass transfer coefficient, respectively; Fig. 9.
L	Latent heat of vaporization of moisture when moisture on the food surface is unfrozen, or for sublimation when it is frozen.

RECTANGULAR COORDINATES - BIDIMENSIONAL HEAT TRANSFER

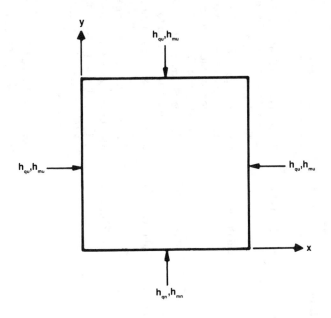

CYLINDRICAL COORDINATES

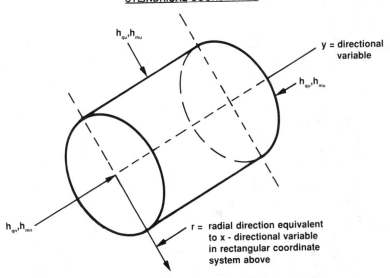

FIG. 9. Illustrating coordinate system for eqns (33)–(35).

TABLE 5
MAXIMAL, MINIMAL AND FREQUENTLY OBSERVED PARAMETERS FOR THE FREEZING AND THAWING OF FOODS

Parameter	Minimal	Frequent	Maximal	Unit	Description
Thermal conductivity					
k_{lx}	3·0	4·1	6·0	cal/(h cm C°)	Thermal conductivity of food at T_{sh} in the x direction (or radial direction for cylinder); eqn (28)
S_{lx}	$1·09 \times 10^{-3}$	$1·05 \times 10^{-2}$	$8·4 \times 10^{-2}$	cal/(h cm C°2)	Rate of change of k as a function of temperature $(T) > T_{sh}$, in the x direction (or radial direction for cylinder); eqn (28)
k_{rx}	8·7	14·3	18·9	cal/(h cm C°)	Reference x-directional thermal conductivity of frozen food; eqn (24)
S_{sx}	−0·05	0·031	0·172	cal/(h cm C°2)	Rate of change of k as a function of $T \leq T_{sh}$, in the x-direction or radial direction for cylinder; eqn (24)
S_{ly}	$1·3 \times 10^{-3}$	$1·52 \times 10^{-2}$	0·110	cal/(h cm C°2)	Rate of change of k as a function of $T > T_{sh} =$ in the y direction; eqn (28)
γ	0·866	0·917	0·958	—	$= k_{ry}/k_{rx}$, where k_{ry} is the reference y-directional k; eqn (24)
Density					
ρ_l	0·88	1·04	1·23	g/cm³	Density of thawed food; eqn (27)
ρ_r	0·82	1·00	1·15	g/cm³	Reference density of frozen food; eqn (25)
S_d	0·00	2×10^{-4}	6×10^{-4}	g/(cm³C°)	Rate of change of ρ as a function of $T \leq T_{sh}$

Effective specific heat

c_1	0·35	0·76	0·97	cal/(g C°)	Specific heat of thawed food; eqn (26)
c_e	0·13	0·34	0·47	cal/(g C°)	Empirical constant for effective specific heat ($T \le T_{sh}$); eqn (23)
D	31·5	82·0	192·2	cal/(g C$^{°1-n}$)	Empirical constant to predict effective specific heat c, $T \le T_{sh}$; eqn (23)
n	1·41	1·78	2·0	—	Empirical constant to predict c; eqn (23)

Surface heat or moisture transfer

h_{qu}	0·43	4·0	68·9	cal/(h cm² C°)	Surface heat transfer coefficient h on all exposed surfaces except bottom surface of rectangle or cylinder during freezing or thawing
h_{qn}	0·43	$\le 4·0$	$\le 68·9$	cal/(h cm² C°)	Value of h on bottom surface
h_{mu}	0·0	$3·6 \times 10^{-4}$	$8·3 \times 10^{-3}$	g/(h cm² kPa)	Convective mass transfer coefficient h_m on all exposed surfaces except bottom
h_{mn}	0·0	$\le 3·6 \times 10^{-4}$	$\le 8·3 \times 10^{-3}$	g/(h cm² kPa)	Value of h_m on bottom surface
ε^1_{qu}	0·0	$3·9 \times 10^{-9}$	$4·83 \times 10^{-9}$	cal/(h cm² C$^{°4}$)	Where $\varepsilon^1 = F\varepsilon\sigma$ in eqn (20), applicable to all exposed surfaces except bottom surface
ε^1_{qn}	0·0	$\le 3·9 \times 10^{-9}$	$\le 4·83 \times 10^{-9}$	cal/(h cm² C$^{°4}$)	Same as above, but applied to bottom surface
$\phi_a = m/m_r^a$	0·2	0·7	1·0	—	Relative humidity of surrounding air during thawing, needed to solve for convective mass transfer coefficient. During freezing it was assumed that $\phi_a = 1·0$

TABLE 5—contd.

Parameter	Minimal	Frequent	Maximal	Unit	Description
Operational constant					
l(rect)	0·3	2·5	15·4	cm	Characteristic dimension of rectangular two-dimensional food, $l = y$-directional side dimension
l(cyl)	0·3	3·5	5·0	cm	Characteristic dimension of finitely cylindrical food, $l =$ radius
T_o (freezing)	−1·0	8·0	30·0	°C	Initial food temperature; uniform throughout the food
(thawing)	−40·0	−20·0	−6·0		
T_a (freezing)	−80·0	−30·0	−10·0	°C	Bulk temperature; eqn (35)
(thawing)	0·0	26·0	263·0		
T_{sh}	−5·6	−2·0	−0·1	°C	Highest freezing temperature of food
T_f (freezing)	−30·0	−21·0	−6·0	°C	Final freezing or thawing temperature at the thermal center of the food
(thawing)	−4·0	12·0	94·0		
T_e (freezing)	$T_a − 15$	$T_a − 7$	$T_a + 5$	°C	Temperature of radiative heat sink
(thawing)	$T_a − 1$	T_a			
S_f(rect)	0·5	4·0	16·0	—	Shape factor of rectangular food, where S_f(rect) $= x$-directional side dimension/l(rect)
S_f(cyl)	0·125	2·4	5·0	—	Shape factor for cylindrical food, where S_f(cyl) $= y$-directional side or height/radius

$^a m =$ Vapor pressure in a space adjacent to the food surface (kPa).

$m_r =$ Vapor pressure evaluated at reference temperature (kPa), which was −40°C for freezing and +8°C for thawing.

m	Partial pressure of water vapor in surrounding atmosphere.
m_r	Value of m at a reference surrounding medium temperature (see footnote, Table 5).
T_a	Freezing or thawing bulk temperatures.
T_e^A	Absolute temperature of heat sink for radiative heat exchange.
ε	Effective emissivity for thermal radiation; eqn (20).

The behavior of the thermophysical properties below and above the highest freezing point were predicted with eqns (23)–(28). The maximal, minimal and frequent values for all the parameters needed to describe freezing or thawing of rectangular two-dimensional slabs or finite cylinders with the use of the equations discussed, are listed in Table 5. Table 6 gives the results of the parametric analyses on these two food shapes when subjected to freezing and thawing. As indicated earlier, for the parametric analyses the heat conduction equations together with thermophysical properties and boundary and operational conditions were transformed into dimensionless expressions. The results of the parametric analyses in Table 6 show which of these dimensionless expressions significantly influenced dimensionless freezing or thawing times. The dimensionless freezing time was defined as:

$$Fo = [k_r/(\rho_r \cdot c_e)]t/l^2 \qquad (36)$$

The terms k_r, ρ_r and c_e for predicting k, ρ and c at $T \leqslant T_{sh}$ with eqns (23), (24) and (25) are defined in Table 5. The physical meaning of the dimensionless groups in Table 6 may be obtained by examining the physical quantities which define them. These physical quantities are given in Table 5. The significant dimensionless groups influencing freezing times of rectangular foods are freezing bulk temperature U_a, highest freezing temperature U_{sh}, final freezing temperature U_f, a constant for estimating effective specific heat C_{eh}, the Biot numbers, Bi_u and Bi_m, and the shape factor S_f. For thawing times of rectangular foods the final thawing temperature U_f was not significant. For cylindrical foods similar results were obtained, except that for freezing the initial food temperature U_o was also significant. According to Hayakawa et al.,[18] differences in the level of significance of the parameters in the two foods could be due to differences in shape factors (Table 5), and differences in thermal responses resulting from their overall configurations.

Of practical interest is that the following parameters did not significantly influence the freezing or thawing rates of the food shapes studied:

—All parameters for estimating temperature dependent anisotropic thermal conductivity and temperature dependent density.

TABLE 6
DIMENSIONLESS TERMS WHICH SIGNIFICANTLY INFLUENCE DIMENSIONLESS FREEZI
AND THAWING TIMES OF RECTANGULAR AND CYLINDRICAL FOOD

Dimensionless group	Definition (refer to Table 5)	Freezing		Thawing	
		Rect-angle	Cyl-inder	Rect-angle	Cyl-inder
U_o	$(T_o - T_{ar})/(T_{or} - T_{ar})$	***	*	***	***
U_a	$(T_a - T_{ar})/(T_{or} - T_{ar}$	*	*	*	*
U_{sh}	$(T_{sh} - T_{ar})/(T_{or} - T_{ar})$	*	*	*	**
U_f	$(T_f - T_{ar})/(T_{or} - T_{ar})$	*	*		
U_e	$(T_e^A - T_{or})/(T_{or} - T_{ar})$				
R_{rx}	$k_{1x}/k_{rx} - 1$				
M_{rx}	$S_{sx}(T_{sh} - T_{ar})/k_{rx}$				
M_{klx}	$S_{1x}(T_{or} - T_{ar})/k_{rx}$				
M_{kly}	$S_{1y}(T_{or} - T_{ar})/k_{ry}$				
γ	k_{ry}/k_{rx}				
R_d	$\rho_1/\rho_r - 1$				
M_d	$S_d(T_{or} - T_{ar})/\rho_r$				
R_c	c_1/c_e				
C_{eh}	$D/[c_e(T_{sw} - T_{sh})^n]$	*	*	*	*
n	See eqn (23)				
Bi_u	$h_{qu} \cdot l/k_{rx}$	*	*	*	*
Bi_m	$h_{qn} \cdot l/k_{ry}$	**	*	**	***
B_{mu}	$h_{mu} \cdot L_r \cdot m_r \cdot l/[k_{rx}(T_{or} - T_{ar})]$				
B_{mn}	$h_{mn} \cdot (L_r/S_f) \cdot m_r \cdot l/[k_{ry}(T_{or} - T_{ar})]$				
NR_u	$F\varepsilon_u\sigma(T_{or} - T_{ar})^3 l/k_{rx}$				
NR_n	$F\varepsilon_n\sigma(l/S_f)(T_{or} - T_{ar})^3/(k_{rx} \cdot S_f)$				
ϕ_a	m/m_r				
S_f	Shape factor; see Notation	*	*	*	*

*Significant influence; level of significance ≤ 0.05.
**Nearly significant influence; $0.05 <$ level of significant ≤ 0.10.
***Moderately significant; $0.10 <$ level of significance as ≤ 0.15.
Note: The dimensionless temperature U was defined as $U = (T - T_{ar})/(T_{or} - T_{ar})$.
For freezing $T_{ar} = -40°C$ and $T_{or} = +8°C$.
For thawing $T_{ar} = +8°C$ and $T_{or} = -40°C$.

—Boundary parameters for determining the radiative heat exchange and moisture loss.

Therefore, the development of simplified equations to predict freezing or thawing times of rectangular and cylindrical foods need only to consider one of the parameters for estimating the effective specific heat, modified Biot numbers, and highest freezing point, in addition to four operational parameters: U_o, U_a, U_{sh} and U_f.

6. PREDICTING HEAT TRANSFER RATES DURING FREEZING AND THAWING USING NUMERIC METHODS

6.1. Introduction

The most reliable and versatile methods for predicting heat transfer with phase change are numerical finite difference and finite element techniques. Since the number of mathematical operations needed to obtain transient state temperatures with these methods is large, a computer is required. However, once the computer program for solving the numerical algorithm is developed, it can be used with ease to predict, usually in a matter of minutes, thermal responses for a variety of conditions which may arise during industrial operations. They are also powerful tools for designing freezers and for optimizing boundary and operational conditions for freezing and thawing. Cleland and Earle[3] assessed freezing time prediction methods, and indicated that numerical methods are the closest to an exact freezing time prediction that exists. The reliability of numerical methods to determine transient state temperatures in foods undergoing phase change has been demonstrated by Hayakawa et al.,[17,18] Succar and Hayakawa,[19] Bonacina et al.,[21] Heldman,[23] Heldman and Gorby,[25] Hohner and Heldman,[26] Hsieh et al.,[27] Bonacina and Comini,[53] Cleland and Earle,[54,55] Nonino and Hayakawa,[58] Succar and Hayakawa,[69] Nonino and Hayakawa,[70] Cleland and Earle,[80,81] Cleland et al.,[82] De Michelis and Calvelo,[83] Joshi and Tao,[84] Katayama and Hattori,[85] Shamsunder and Sparrow,[86,87] Comini et al.,[88] Comini and Del Giudice[89] and Rebellato et al.[90]

Many foods subjected to freezing or thawing may be roughly approximated by simple geometrical shapes: infinite slabs, infinite cylinders, spheres, rectangular packages subjected to two-dimensional heat transfer, brick-shaped foods undergoing three-dimensional heat transfer, and finite cylinders. Finite elements and finite differences have been used to solve the heat conduction equation with phase change for these configurations. For

foods with irregular shapes finite elements are the method of choice. For
foods of simple geometrical shapes finite differences are easier to program
and should be used. In these shapes finite element methods have no
advantages over finite differences.[3] Furthermore, Carnahan et al.[91] discuss
a procedure through which finite differences can be applied to products
with irregular boundaries. This procedure has been successfully used by
Tandon and Bhowmik[92] to predict transient state temperatures in pouches
filled with conduction heating foods.

The sections which follow will be used to describe finite difference
solutions to the heat transfer equation for freezing and defrosting of foods
whose exposed boundaries are subjected to non-symmetric convective and/
or radiative heat exchange. To simplify the description of the numerical
algorithm, the shape of the food assumed is that of an infinite slab, infinite
cylinder or sphere (one-dimensional heat transfer). However, the principles
used can be applied with ease for two- and three-dimensional heat transfer
solutions.

6.2. The Finite Difference Approach

Numerical procedures for the solution of the Fourier heat conduction
equation fall into two categories, implicit and explicit.[91,93-96] Cleland and
Earle[3] presented a thorough critical review of explicit and implicit finite
difference schemes commonly used in freezing time calculations. Explicit
schemes are easier to program and require less computer memory than
implicit schemes. However, the number of calculations required to
determine a freezing or thawing time is very large compared to the implicit
methods. The problem is due to a restrictive stability criterion characteristic
to explicit schemes which is described later. Consider the one-dimensional
form of eqn (13), applicable to infinite slabs (Fig. 10):

$$\rho c \frac{\partial T}{\partial t} = \frac{\partial}{\partial x}\left(k \frac{\partial T}{\partial x}\right) \tag{37}$$

Initial condition:

$$T = T_o \text{ for all } x \text{ at } t = 0 \tag{38}$$

Boundary conditions:

$$k \frac{\partial T}{\partial x} = -h_o(T - T_a) \quad \text{for } x = 0 \text{ and } t > 0 \tag{39}$$

$$-k \frac{\partial T}{\partial x} = -h_l(T - T_a) \quad \text{for } x = l \text{ and } t > 0 \tag{40}$$

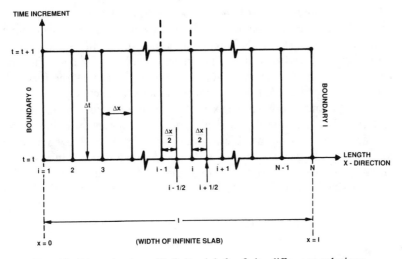

FIG. 10. Discretization of infinite slab for finite difference solutions.

Using Taylor's series expansion, the time and space derivatives in eqns (37), (39) and (40) may be approximated with the following finite difference formulae:

$$\rho_i c_i \frac{\partial T}{\partial t} \cong \rho_i^t c_i^t \frac{T_i^{t+1} - T_i^t}{\Delta t} \tag{41}$$

$$\frac{\partial T}{\partial x} \cong \frac{T_{i+1}^t - T_{i-1}^t}{2\Delta x} \tag{42}$$

$$\frac{\partial T}{\partial x} \cong \frac{T_i^t - T_{i-1}^t}{\Delta x} \tag{43}$$

$$\frac{\partial}{\partial x}\left(k\frac{\partial T}{\partial x}\right) \cong \frac{k_{i-(1/2)}^t T_{i-1}^t - 2k_i^t T_i^t + k_{i+(1/2)}^t T_{i+1}^t}{(\Delta x)^2} \tag{44}$$

Subscript i refers to a node in the grid-point system shown in Fig. 10. The grid is obtained by subdividing the food's length (in the x-direction in this case) in $N-1$ equal parts. This subdivision or discretization delivers N nodes or points at which temperatures will be evaluated. The space increment Δx is defined as

$$\Delta x = l/(N-1) \tag{45}$$

The superscript t in eqns (41)–(44) describes the position of the solution in time (Fig. 10); T^t represents the temperatures evaluated at the current time step in node i, and T^{t+1} is the unknown temperature in this same node at a future or advanced time step. The time steps advance at constant time increments Δt. Equations (41), (42) and (43) are known as the forward, central and backward difference approximations, respectively.[94,97]

Small differences between the exact derivatives on the left hand sides of these equations and their numerical approximations in their right hand sides, are due to truncation of the Taylor's series expansion. However, these truncation errors become smaller as Δt and Δx tend to zero. Substituting eqns (41)–(44) in eqns (37)–(40), and rearranging, delivers a form of the Euler explicit finite difference scheme.[95,96]

$$T_i^{t+1} = T_i^t + \frac{\Delta t}{(\Delta x)^2} \left[\alpha_{i-(1/2)}^t T_{i-1}^t - (\alpha_{i-(1/2)}^t + \alpha_{i+(1/2)}^t) T_i^t + \alpha_{i+(1/2)}^t T_{i+1}^t \right] \quad (46)$$

Initial condition:

$$T_i^t = T_o \text{ for } i = 1, 2, 3, \ldots, N \quad (47)$$

Boundary conditions:

$$T_1^{t+1} = \gamma_{1+(1/2)}^t T_2^{t+1} + \Delta x T_a h_o \gamma_{1+(1/2)}^t / k_{1+(1/2)}^t \quad (48)$$

$$T_N^{t+1} = \beta_{N-(1/2)}^t T_{N-1}^{t+1} + \Delta x T_a h_1 \beta_{N-(1/2)}^t / k_{N-(1/2)}^t \quad (49)$$

where:

$$\alpha_{i\pm(1/2)} = k_{i\pm(1/2)} / (\rho_i c_i) \quad (50)$$

$$\gamma_{1+(1/2)} = 1 / (1 + \Delta x h_o / k_{1+(1/2)}) \quad (51)$$

$$\beta_{N-(1/2)} = 1 / (1 + \Delta x h_1 / k_{N-(1/2)}) \quad (52)$$

To solve eqns (48) and (49) one must first solve eqn (46), for nodes $i = 2$ to $i = N - 1$. Analyses of eqns (46)–(52) shows that in the explicit numerical scheme temperatures in the future time step are determined from known temperatures and thermophysical properties at the present time step. The thermophysical properties at the nodal points or their arithmetic averages between nodal points (Fig. 10) may be obtained with eqns (24)–(28), except for c_i which should be determined as follows:

$$c_i = (c_{i+(1/2)} + c_{i-(1/2)}) / 2 \quad (53)$$

where

$$c_{i\pm(1/2)} = (H_i - H_{i\pm1}) / (T_i - T_{i\pm1}) \quad (54)$$

The enthalpy value H may be predicted with eqn (29) or eqn (30). This procedure to approximate the effective specific heat is necessary to maintain a proper heat balance while applying the numerical sheme.[17] Attempts to estimate c_i with eqn (23) or (26) should be avoided since when the food material reaches its freezing point in between two nodal points neither equation will detect the sharp effective specific heat peak (Fig. 4), and freezing or thawing times will be underestimated.

In the case of heat conduction with no phase change, the thermal diffusivity α (eqn 50) may be assumed constant. Vichnevetsky[96] used the von Neumann analysis of numerical stability to demonstrate that the explicit scheme is stable if it satisfies the following condition:

$$\alpha \Delta t / (\Delta x)^2 \leqslant \tfrac{1}{2} \tag{55}$$

The term stability describes the numerical behavior of the solutions by a finite difference scheme with advancing time-marching increments. A stable scheme is one in which the conditions for the numerical solutions are such that the magnitude of oscillation of these solutions around the true solutions are prevented from growing exponentially in the time-marching calculations. For the cases of freezing and thawing, this stability condition given by eqn (55) must be satisfied at any time step or nodal point during the numerical solution. Therefore the k, ρ and c values of the fully frozen food should be used as the reference to satisfy the stability criteria, since these values will deliver the largest $\alpha \Delta t / (\Delta x)^2$. The α value of the fully frozen food can be roughly 8–10 times larger than that of thawed foods. This characteristic imposes a serious restriction on the size of Δt. For example, for Karlsruhe test material with thermophysical properties similar to beef meat, the values of k, ρ and c at $-40°C$ are $1·7$ W/(m C°), 945 kg/m^3, and $0·339$ kcal/(kg C°), respectively,[18,19] Assuming a typical product thickness of $0·025$ m which is subdivided into 10 equal parts ($\Delta x = 0·0025$), the time increment should be:

$$\Delta t \leqslant 6·8 \times 10^{-4} \text{ hours} \tag{56}$$

This means that if a freezing process requires 1 h for completion, the numerical scheme would require 1470 time steps to reach the solution. This undesirable restriction in the spatial or time step sizes that can be used, is eliminated by using the fully implicit finite difference technique which is unconditionally stable.[96] The implicit method overcomes these difficulties at the expense of a somewhat more complicated calculational procedure, since the algorithm requires the solution of a non-linear system of simultaneous equations. However, Carnahan et al.[91] state that the absence of restriction on the size $\Delta t/(\Delta x)^2$ in the implicit method usually outweighs this moderate increase in computational effort.

6.3. The General Implicit Finite Difference Technique

The general fully implicit finite difference technique was used by Succar and Hayakawa[19,20,69] to solve the heat transfer equation with phase change for foods whose shape could be approximated by infinite slabs, infinite cylinders and spheres. For the case of the infinite slab, their

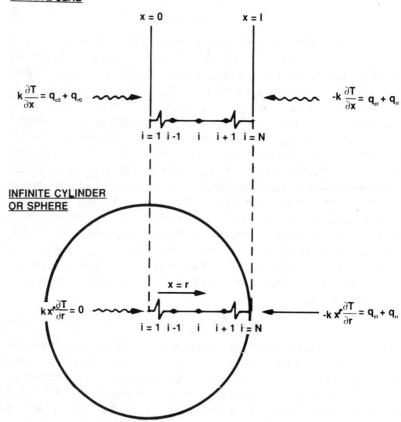

INFINITE SLAB

INFINITE CYLINDER OR SPHERE

FIG. 11. Cross-sectional views of infinite slabs or infinite cylinders and spheres, showing discretization and boundary conditions for finite difference solution of partial differential equation for heat transfer.

solutions assumed non-symmetric convective and radiative heat exchange on the boundaries, as illustrated in Fig. 3.

The following non-linear parabolic equation was used to define transient state heat transfer in infinite slabs, infinite cylinders and spheres:[11]

$$x^p \rho c \frac{\partial T}{\partial t} = \frac{\partial}{\partial x}\left(x^p k \frac{\partial T}{\partial x}\right) \tag{57}$$

for all x except on the boundaries and for $t > 0$; $p = 0$ for infinite slabs, $p = 1$ for infinite cylinders, and $p = 2$ for spheres. Also note that x becomes a radial directional variable for cylindrical and spherical configurations (Fig. 11).

Initial condition:

$$T = T_0 \text{ at } t = 0, \text{ for } 0 \leqslant x \leqslant l \tag{58}$$

Boundary conditions (Fig. 11):

At boundary $x = l$, surrounding infinite cylinders or spheres, or on the right hand side of infinite slabs:

$$-x^p k \left.\frac{\partial T}{\partial x}\right]_{x=l} = -x^p h_1(T - T_a) - x^p F \varepsilon_1 \sigma[(T^A)^4 - (T_{e1}^A)^4] \tag{59}$$

At boundary, $x = 0$, applicable to infinite slabs:

$$x^p k \left.\frac{\partial T}{\partial x}\right]_{x=0} = -x^p h_o(T - T_a) - x^p F \varepsilon_o \sigma[(T^A)^4 - (T_{eo}^A)^4] \tag{60}$$

and for infinite cylinders or spheres:

$$x^p k \left.\frac{\partial T}{\partial x}\right]_{x=0} = 0 \tag{61}$$

The terms in eqns (59) and (60) have been defined in eqns (21) and (22). Equations (59) and (60) applied to the exposed surfaces of infinite slabs assume that there are non-symmetric convective and radiative heat exchange on the boundaries. The thermophysical properties k, ρ and c appearing in eqns (57)–(60) may be predicted with eqns (23)–(28).

Equation (57) is approximated by a general implicit finite difference technique as follows:

$$\frac{T_i^{t+1} - T_i^t}{\Delta t} = \frac{(1-\theta)}{(\Delta x)^2} \cdot \frac{1}{V_i^t}\left(\eta_{i-(1/2)}^t T_{i-1}^t - \gamma_i^t T_i^t + \eta_{i+(1/2)}^t T_{i+1}^t\right)$$

$$+ \frac{\theta}{(\Delta x)^2} \cdot \frac{1}{V_i^{t+1}}\left(\eta_{i-(1/2)}^{t+1} T_{i-1}^{t+1} - \gamma_i^{t+1} T_i^{t+1} + \eta_{i+(1/2)}^{t+1} T_{i+1}^{t+1}\right) \tag{62}$$

where T_i^{t+1} is the food temperature at the advanced time step $t+1$ at location i, as shown in Fig. 10. Other terms in eqn (62) are:

$$V_i = x_i^p \rho_i c_i; \tag{63}$$

$$\eta_{i \pm (1/2)} = x_{i \pm (1/2)}^p k_{i \pm (1/2)} \tag{64}$$

$$\gamma_i = x_i^p (k_{i-(1/2)} + k_{i+(1/2)}) \tag{65}$$

As discussed previously for the estimation of c_i in eqn (63), it is necessary to use eqn (53); k and ρ may be predicted with eqns (24) and (25) for $T \le T_{sh}$, or eqns (27) and (28) for $T > T_{sh}$.

The term θ may assume any value in the range $\frac{1}{2} \le \theta < 1$. For $\theta = 0$, eqn (62) becomes the explicit finite difference approach described with eqn (46).

The initial condition becomes:

$$T_i^t = T_o \text{ for } i = 1, N \text{ at time } t = 0. \tag{66}$$

The boundary conditions become (Figs 3, 10 or 11):
At $x = l(i = N)$:

$$-x_{N-(1/2)}^p k_{N-(1/2)}^{t+1} \frac{T_{N-1}^{t+1} - T_N^{t+1}}{\Delta x} = -x_{N-(1/2)}^p h_l(T_N^{t+1} - T_a)$$

$$-x_{N-(1/2)}^p F \varepsilon_l \sigma [(T_N^{t+1} + T_{abs})^4 - (T_{el} + T_{abs})^4] \tag{67}$$

In eqn (67) T_{abs} = absolute temperature.
At $x = 0$, $(i = 1)$:

$$x_{1+(1/2)}^p k_{1+(1/2)}^{T+1} \frac{T_1^{T+1} - T_2^{T+1}}{\Delta x} = -x_{1+(1/2)}^p h_o(T_1^{t+1} - T_a)$$

$$-x_{1+(1/2)}^p F \varepsilon_o \sigma [(T_1^{t+1} + T_{abs})^4 - (T_{eo} + T_{abs})^4] \tag{68}$$

Equation (68) is also applicable to infinite cylinders and spheres, provided that

$$h_o = 0 \tag{69}$$

$$\varepsilon_o = 0 \tag{70}$$

Final freezing or thawing times are those at which the following condition is satisfied:

$$T_c^{t+1} = T_f \tag{71}$$

where T_c is the temperature at the thermal center and T_f is the prescribed final freezing or thawing temperature.

Examination of eqns (62), (67) and (68) indicates that in the general implicit finite difference scheme, the temperatures at an advanced time step are derived from unknown temperatures and thermophysical properties at this advanced time, as well as those evaluated at the current time step (known values). Due to this characteristic the general implicit schemes deliver a more accurate representation of the heat transfer phenomena.

Vichnevetsky[96] shows that general implicit schemes are consistent and unconditionally stable for $\frac{1}{2} \leqslant \theta < 1$. Values of $\theta = \frac{1}{2}$ assign equal emphasis to temperatures and thermophysical properties at both the current and advanced time steps. When $\theta = \frac{1}{2}$, the scheme is called the Crank–Nicolson method.[97] The term stability has been previously discussed. The term consistency mentioned above means that the numerical procedure approximates the true solution of the partial differential equation. Carnahan et al.[91] discussed that unless precautions are taken when selecting Δx and Δt, certain explicit approximations (i.e. an approximation proposed by DuFort and Frankel[98]) will not be consistent with the Fourier heat conduction equation. Another advantage of the general implicit scheme proposed is that it converges with smaller truncation errors than explicit schemes.

6.3.1. Solving the fully implicit algorithm
When all unknowns in eqns (62)–(68) (i.e. terms with superscript $t + 1$) are placed on the left hand side, the following equations are obtained.

For nodal points $i = 2$ through $N - 1$, eqn (62):

$$-\mu_{i-(1/2)}^{t+1} T_{i-1}^{t+1} + \psi_i^{t+1} T_i^{t+1} - \mu_{i+(1/2)}^{t+1} T_{i+1}^{t+1} = D_i^t \tag{72}$$

where

$$D_i^t = T_i^t + \left(\frac{1}{\theta} - 1\right)[\mu_{i-(1/2)}^t T_{i-1}^t - (\psi_i^t - 1)T_i^T + \mu_{i+(1/2)}^t T_{i+1}^t] \tag{73}$$

$$\mu_{i\pm(1/2)} = \omega \eta_{i\pm(1/2)}/V_i \tag{74}$$

$$\psi_i = 1 + \omega \gamma_i / V_i \tag{75}$$

$$\omega = \theta \Delta t/(\Delta x)^2 \tag{76}$$

For nodal point $i = N$ (right hand side boundary), eqn (67):

$$-\eta_{N-(1/2)}^{t+1} T_{N-1}^{t+1} + \lambda_{N-(1/2)}^{t+1} T_N^{t+1} = z_{N-(1/2)}(h_l T_a - F\varepsilon_l \sigma[(T_N^{t+1} + T_{abs})^4$$
$$-(T_{el} + T_{abs})^4])$$

$$= D_N^{t+1} \qquad (77)$$

For nodal point $i = 1$ (left hand side boundary), eqn (68):

$$\lambda_{i+(1/2)}^{t+1} T_1^{t+1} - \eta_{1+(1/2)}^{t+1} T_2^{t+1} = z_{1+(1/2)}(h_o T_a - F\varepsilon_o \sigma[(T_1^{t+1} + T_{abs})^4$$
$$-(T_{e\sigma} + T_{abs})^4])$$

$$= D_1^{t+1} \qquad (78)$$

where:

$$z_{i\pm(1/2)} = x_{i\pm(1/2)}^p \Delta x \qquad (79)$$

$$\lambda_{i\pm(1/2)} = z_{i\pm(1/2)} h_x + \eta_{i+(1/2)} \qquad (80)$$

$$h_x = h_o \text{ for } i = 1 \qquad (81)$$

$$h_x = h_i \text{ for } i = N \qquad (82)$$

For infinite cylinders and spheres we may assume that $h_o = \varepsilon_o = 0$, in eqn (78); in these cases $D_1^{t+1} = 0$ (Fig. 11).

Equations (72), (77) and (78) may now be assembled in the form of simultaneous equations describing food temperatures at each of the N nodal points selected (Figs 10 or 11). In the matrix form such an assembly results in the following equation:

$$[\vec{\vec{A}}^{t+1}][\vec{T}^{t+1}] = [\vec{D}] \qquad (83)$$

Our objective is to obtain the solution for the column vector $[\vec{T}^{t+1}]$, namely the food temperatures at the N nodal points in the future or advanced time step.

The matrix $[\vec{\vec{A}}^{t+1}]$ is a matrix of variable coefficients. These coefficients are related to the thermophysical properties at the advanced time step (i.e. they depend on T^{t+1}). This matrix is a tridiagonal matrix as shown opposite:

$$[\bar{\bar{A}}^{t+1}] = \begin{bmatrix} \lambda^{t+1}_{1+(1+2)} & -\eta^{t+1}_{1+(1/2)} & 0 & 0 & & & & & \cdot \\ -\mu^{t+1}_{2-(1/2)} & \psi^{t+1}_2 & -\mu^{t+1}_{2+(1/2)} & 0 & & & & & \cdot \\ & -\mu^{t+1}_{3-(1/2)} & \psi^{t+1}_3 & -\mu^{t+1}_{3+(1/2)} & 0 & & & & \cdot \\ & & -\mu^{t+1}_{4-(1/2)} & \psi^{t+1}_4 & -\mu^{t+1}_{4+(1/2)} & & & & \cdot \\ & & & \ddots & & & & & \cdot \\ & & & & & -\mu^{t+1}_{(N-2)-(1/2)} & \psi^{t+1}_{N-2} & -\mu^{t+1}_{(N-2)+(1/2)} & 0 \\ & & & & & 0 & -\mu^{t+1}_{(N-1)-(1/2)} & \psi^{t+1}_{(N-1)} & -\mu^{t+1}_{(N-1)+(1/2)} \\ & & & & & 0 & 0 & -\eta^{t+1}_{N-(1/2)} & \lambda^{t+1}_{N-(1/2)} \end{bmatrix} \tag{84}$$

$$[\vec{T}^{t+1}] = \begin{bmatrix} T_1^{t+1} \\ T_2^{t+1} \\ T_3^{t+1} \\ \cdot \\ \cdot \\ \cdot \\ T_{N-1}^{t+1} \\ T_N^{t+1} \end{bmatrix} \tag{85}$$

The resultant vector $[\vec{D}]$ is defined by:

$$[\vec{D}] = \begin{bmatrix} D_1^{t+1} \\ D_2^{t} \\ D_3^{t} \\ \cdot \\ \cdot \\ \cdot \\ D_{N-1}^{t} \\ D_N^{t+1} \end{bmatrix} \tag{86}$$

Notice that for infinite slabs, the first and last elements of $[\vec{D}]$ are unknown functions of T^{t+1}, given by the right hand sides of eqns (77) and (78). For infinite cylinders and spheres only D_N^{t+1} is unknown and $D_1^{t+1} = 0$.

Solving for $[\vec{T}^{t+1}]$ in eqn (83) does not require inverting matrix $[\vec{\vec{A}}^{t+1}]$. Instead it can be easily solved by a Gaussian elimination method described by Carnahan *et al.*[91] Since matrix $[\vec{\vec{A}}^{t+1}]$ and vector $[\vec{D}]$ in eqn (83) are functions of T^{t+1}, the temperature values T_i^{t+1} for all i's may be approximated through the Jacobi (or simultaneous displacements) iteration procedure. In this procedure we arbitrarily choose a starting value (initial guess) for all T^{t+1} values, namely $[\vec{T}^{t+1}]^k$, where superscript k is the order of the approximate solution.

For the first time step in which we need to estimate $[\vec{T}^{t+1}]$ at time $t = 0 + \Delta t$, our initial guess becomes the initial condition given by eqn (66), namely $[\vec{T}^{t+1}]^k = [\vec{T}_0]$. With these starting values, the coefficients in matrix $[\vec{A}^{t+1}]^k$ and in the resultant vector $[\vec{D}]$ are estimated. We then solve for $[\vec{T}^{t+1}]$ using the Gaussian elimination method[91] to obtain $[\vec{T}^{t+1}]^{k+1}$. By making this newly found $(k+1)$st solution vector our new guess,

$$[\vec{T}^{t+1}]^k = [\vec{T}^{t+1}]^{k+1} \tag{87}$$

and by redetermining matrix $[\vec{A}^{t+1}]$ and vector $[\vec{D}]$, the procedure is repeated until $[\vec{T}^{t+1}]^{k+1}$ and $[\vec{T}^{t+1}]^k$ converge, when the following condition is satisfied:

$$[\vec{T}^{t+1}]^{k+1} - [\vec{T}^{t+1}]^k \leqslant E \approx 0 \tag{88}$$

The value of E, Δt and Δx are the choice of the user. Smaller E, Δt and Δx values will deliver more accurate representation of the true solution of the heat conduction equation. The value of θ should be in the range $0.5 \leqslant \theta < 1$, which as indicated earlier makes the general implicit numerical scheme unconditionally stable.[96] Succar and Hayakawa[19,20,69] obtained good results using $E = 1 \times 10^{-3}$, $\Delta x = l/25$, and θ values of 0.75 or 0.5. The choice of Δx was found to have a significant effect on the value of Δt which can be used. By comparing numerical solutions with varying Δx and Δt values to experimental data, the following criteria for the selection of $\Delta t/(\Delta x)^2$ was found:

(a) For $l/25 \leqslant \Delta x < l/15$:

$$0.25 \leqslant \alpha_r \Delta t/(\Delta x)^2 \leqslant 100 \tag{89}$$

(b) For $l/15 \leqslant \Delta x \leqslant l/10$:

$$0.25 \leqslant \alpha_r \Delta t/(\Delta x)^2 \leqslant 10 \tag{90}$$

The value of α_r is a reference thermal diffusivity defined by:

$$\alpha_r = k_r/(\rho_r c_e) \tag{91}$$

Where c_e, k_r and ρ_r are empirical coefficients in eqns (23), (24) and (25), respectively. Values of c_e, k_r and ρ_r for a variety of food products are listed in Tables 1, 2 and 3, respectively.

When Δt values were chosen following the criteria defined by eqns (89) and (90), differences between experimental freezing times and computed freezing times were 3% or less. If the criteria were not followed, accumulated errors due to the truncation of Taylor's series expansion, needed to approximate the partial derivations of the heat conduction equation, became significantly large, resulting in prediction errors as large as 30%. Typical transient state temperatures at the thermal center of spheres and infinite cylinders, obtained through a computer package which uses the general implicit procedure discussed, are shown in Figs 12 and 13. In these figures agreements between experimental and computed freezing times are very good. Similar results were obtained during freezing and thawing of infinite slab-shaped samples prepared with Karlsruhe test material.[69] The coefficients for convective and radiative heat transfer needed for these computations were obtained using a response surface procedure developed by Succar and Hayakawa.[20,69]

Figure 12 shows the three main periods which occur during freezing of foods.[99] A fourth period is the supercooling period[68,100] which is not accounted for with the mathematical model discussed. However, according to Bakal and Hayakawa,[16] the duration of the supercooling period is very short and in many cases does not appear at all during freezing experiments.

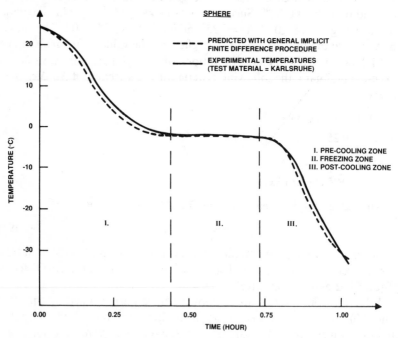

FIG. 12. Predicted versus experimental temperatures during freezing of spheres (measured at the geometrical center).

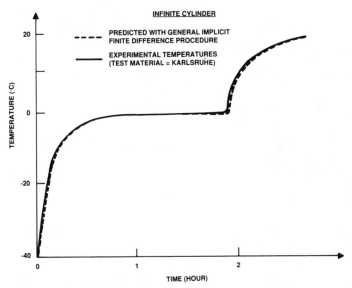

FIG. 13. Predicted versus experimental temperatures during thawing of infinite cylinders (measured at the geometrical center).

7. CONCLUSIONS

Optimizing the design and operation of equipment for freezing and defrosting is becoming of vital importance in this competitive industrial field. Significant amounts of energy may be lost due to improper design and handling. Generally the food engineer has relied on available theoretical or semi-theoretical models to size industrial equipment. With the widespread use of personal computers it is now possible to use numerical procedures as the most reliable tool for optimizing the design and operation of freezing and defrosting equipment. For this purpose it is necessary for the food engineer to understand the application of numerical finite difference or finite element techniques, and the fundamental principles which govern the phenomena of heat transfer in foods undergoing freezing and thawing.

REFERENCES

1. PLANK, R. Z. Ges. Kalte-Ind. Bieh., 1941, 3(10), 1.
2. GOODMAN, T. R. J. Heat Transfer, Transactions ASME, 1958, Series C80:335.

3. CLELAND, A. C. and EARLE, R. L. *J. Food Sci.*, 1984, **49**, 1034.
4. MOORE, W. J. *Physcial Chemistry*, 3rd edn, Prentice-Hall, Inc., Englewood Cliffs, New Jersey, 1962.
5. HELDMAN, D. R. and SINGH, R. P. *Food Process Engineering*, 2nd edn, AVI Publishing Co., Westport, CT, 1981.
6. BARTLETT, L. H. *Refrig. Eng.*, 1944, **47**, 377.
7. CHEN, C. S. *J. Food Sci.*, 1985, **50**(5), 1158.
8. CHEN, C. S. *J. Food Sci.*, 1985, **50**(5), 1163.
9. CHEN, C. S. *J. Food Sci.*, 1986, **51**(6), 1537.
10. CHEN, C. S. *J. Food Sci.*, 1987, **52**(2), 433.
11. CARSLAW, H. S. and JAEGER, J. C. *Conduction of Heat in Solids*, 2nd edn, Oxford University Press, London, 1959.
12. LUIKOV, A. V. *Analytical Heat Diffusion Theory*, Academic Press, New York, 1968.
13. HOLMAN, J. P. *Heat Transfer*, 5th edn, McGraw-Hill Book Company, New York, 1981.
14. FAHIEN, R. W. *Fundamentals of Transport Phenomena*, McGraw-Hill Book Company, New York, 1983.
15. HAYAKAWA, K. *AICHE, Modular Instruction Series, C4.5*, 1985, **4**, 29.
16. BAKAL, A. and HAYAKAWA, K. *Adv. Food Res.*, 1973, **20**, 217.
17. HAYAKAWA, K., NONINO, C., SUGGAR, J., COMINI, G. and DEL GIUDICE, S. *J. Food Sci.*, 1983, **48**(6), 1849.
18. HAYAKAWA, K., NONINO, C. and SUCCAR, J. *J. Food Sci.*, 1983, **48**(6), 1841.
19. SUCCAR, J. and HAYAKAWA, K. Ph.D. thesis, 1984, Rutgers State University, USA.
20. SUCCAR, J. and HAYAKAWA, K. *J. Food Sci.*, 1986, **51**(5), 1314.
21. BONACINA, C., COMINI, G., FASANO, A. and PRIMICERIO, M. *Int. J. Heat Mass Transfer*, 1974, **17**, 861.
22. SUCCAR, J. *ASHRAE Transactions*, 1985, **91**(2), 312.
23. HELDMAN, D. R. *ASAE Transactions*, 1974, **17**(1), 63.
24. HELDMAN, D. R. *Food Technol.*, 1982, **36**(2), 92.
25. HELDMAN, D. R. and GORBY, D. P. *ASAE Transactions*, 1975, **18**, 156.
26. HOHNER, G. A. and HELDMAN, D. R. Computer simulation of freezing rates in foods. Presented at 30th Annual Meeting, Inst. of Food Technologists, San Francisco, CA, 1970.
27. HSIEH, R. C., LEREW, L. E. and HELDMAN, D. R. *Proc. Eng.*, 1977, **1**, 183.
28 LESCANO, C. E. and HELDMAN, D. R. Freezing rates in codfish muscle. Paper No. 73:367 presented at the American Society of Agricultural Engineers, St. Joseph, Michigan, 1973.
29. SCHWARTZBERG, H. G. *J. Food Sci.*, 1976, **41**, 152.
30. SCHWARTZBERG, H. G. *Refrigeration Science and Technology*, International Institute of Refrigeration, Commissions C1 and C2, 1977, p. 303.
31. SCHWARTZBERG, H. G., ROSENAU, J. R. and HAIGHT, J. R. *Refrigeration Science and Technology*, International Institute of Refrigeration, Commissions C1 and C2, 1977, p. 311.
32. LEVY, F. L. *Int. J. Refrig.*, 1982, **5**(3), 155.
33. LEVY, F. L. *Lebensmittle Technol. und Verfahrenstechnik*, 1982, **33**(6), 455.
34. LEVY, F. L. *Int. J. Refrig.*, 1981, **4**(4), 223.

35. JASON, A. C. and LONG, R. A. K. *Int. Congr. Refrig. Proc.*, 9th Congr. Paris, 1955, **1**, 2160.
36. SUCCAR, J. and HAYAKAWA, K. *Lebensm.-Wiss. u.-Technol.*, 1983, **16**, 326.
37. RIEDEL, L. *Chem. Mikrobiol, Technol. Lebenson.*, 1978, **5**, 129.
38. CHANG, H. D. and TAO, L. C. *J. Food Sci.*, 1981, **46**, 1493.
39. RAMASWAMY, H. S. and TUNG, M. A. *J. Food Sci.*, 1981, **46**, 724.
40. METZLER, C. M., ELFRING, G. L. and McEWEN, A. J. *Research Biostatistics*, The Upjohn Company, Kalamazoo, Michigan, 1976.
41. SCHWARTZBERG, H. G. 1985, personal communication.
42. RIEDEL, L. *Kaltetechnik*, 1956, **8**, 374.
43. RIEDEL, L. *Kaltetechnik*, 1957, **9**, 38.
44. RIEDEL, L. *Refrig. Eng.*, 1951, **59**, 670.
45. RIEDEL, L. *Kaltetechnik*, 1960, **12**, 222.
46. MULLER, V. C. F. *Kaltetechnik*, 1954, **6**, 156.
47. SUCCAR, J. and HAYAKAWA, K. *J. Food Sci.*, 1989, submitted for publication.
48. LENTZ, C. P. *Food Technol.*, 1961, **15**, 243.
49. ASHRAE. *Handbook-Fundamentals*, Am. Soc. Htg Refrig. and Air Cond. Engrs, New York, 1981, Chapter 31.
50. SASTRY, S. K. and DATTA, A. K. *Can. Inst. Food Sci. Technol.*, 1984, **17**(4), 242.
51. DICKERSON, R. W., Jr. *The Freezing Preservation of Foods*, 4th edn, Vol. 2, AVI Publishing Co., Westport, CT, 1968.
52. PERRY, R. H. and CHILTON, C. H. *Chemical Engineers Handbook*, 5th edn, McGraw-Hill Book Company, New York, 1973.
53. BONACINA, C. and COMINI, G. *Int. J. Heat Mass Transfer*, 1973, **16**, 581.
54. CLELAND, A. C. and EARLE, R. L. *Int. J. Heat Mass Transfer*, 1977, **20**, 1029.
55. CLELAND, A. C. and EARLE, R. L. *J. Food Sci.*, 1979, **44**, 958.
56. HAYAKAWA, K. I.I.R., commissions C1, C2, 1977, p. 293.
57. HAYAKAWA, K., SCOTT, K. R. and SUCCAR, J. *ASHRAE Transactions*, 1985, **91**(2), 371.
58. NONINO, C. and HAYAKAWA, K. *J. Food Sci.*, 1986, **51**, 116.
59. WALLHOEVER, K., KOERBER, C., SCHEIVE, M. W. and HARTMANN, U. *Int. J. Heat Mass Transfer*, 1985, **28**, 761.
60. ILICALI, C. and SAGLAM, N. *J. Food Proc. Eng.*, 1987, **9**, 299.
61. CHAN, S. H., CHO, D. H. and KOCAMUSTAFAOGULLARI, G. *Int. J. Heat Mass Transfer*, 1986, **26**, 621.
62. PHAM, Q. T. *J. Food Tech.*, 1986, **21**, 209.
63. PHAM, Q. T. *J. Food Sci.*, 1987, **52**(3), 795.
64. DE CINDIO, B., IORIO, G. and ROMANO, V. *J. Food Sci.*, 1985, **50**, 1463.
65. TAOUKIS, P., DAVIS, E. A., DAVIS, H. T., GORDON, J. and TALMON, Y. *J. Food Sci.*, 1987, **52**(2), 455.
66. LACROIX, C. and CASTAIGNE, F. *Can. Inst. Food Sci. Technol. J.*, 1978, **20**(4), 251.
67. RAMASWAMY, H. S. and TUNG, M. A. *J. Food Proc. Eng.*, 1984, **7**, 169.
68. HOLDSWORTH, D. S. In: *Developments in Food Preservation* (ed. S. Thorne), Vol. 4, Elsevier Applied Science Publishers, London, 1987, p. 153.
69. SUCCAR, J. and HAYAKAWA, K. *J. Food Sci.*, 1984, **49**(2), 468.
70. NONINO, C. and HAYAKAWA, K. Freezing or Thawing Time Estimation of Spheroidal Food, 1985, personal communication.

71. CHAPMAN, A. J. *Heat Transfer*, 2nd edn, The MacMillan Co., New York, 1967.
72. PLACKETT, R. L. and BURMAN, J. P. *Biometrika*, 1946, **33**, 305.
73. KLEIJNEN, J. P. C. *Technometrics*, 1975, **17**, 487.
74. RIEDEL, L. *Kaltetechnik*, 1950, **2**, 201.
75. TRESSLER, D. K., VAN ARSDEL, W. B. and COPLEY, M. J. (eds). *The Freezing Preservation of Foods*, Vols 1–4, AVI Publishing Co., Westport, CT, 1968.
76. BARR, A. J., GOODNIGHT, J. H., SALL, J. P. and HELWIG, J. Y. *A User's Guide to SAS*, SAS Institute, Inc., Raleigh, NC, 1976.
77. COCHRAN, W. G. and COX, G. M. *Experimental Designs*, 2nd edn, John Wiley and Sons, New York, 1957.
78. DAVIES, O. L. *The Design and Analysis of Industrial Experiments*, Imperial Chemical Industries, London, 1978.
79. BOX, G. E. P., HUNTER, W. G. and STUART, H. J. *Statistics for Experimenters*, John Wiley and Sons, New York, 1981.
80. CLELAND, A. C. and EARLE, R. L. *J. Food Sci.*, 1977, **42**, 1390.
81. CLELAND, A. C. and EARLE, R. L. *J. Food Sci.*, 1979, **44**, 964.
82. CLELAND, A. C., EARLE, R. L., and CLELAND, D. J. *Int. J. Refrig.*, 1982, **5**, 294.
83. DE MICHELIS, A. and CALVELO, A. *J. Food Sci.*, 1983, **48**, 909.
84. JOSHI, C. and TAO, L. C. *J. Food Sci.*, 1974, **39**, 623.
85. KATAYAMA, K. and HATTORI, M. *JSME Bulletin*, 1975, **18**, 41.
86. SHAMSUNDER, N. and SPARROW, E. M. *J. Heat Transfer, ASME Transactions*, Series C, 1975, **97**, 333.
87. SHAMSUNDER, N. and SPARROW, E. M. *J. Heat Transfer, ASME Transactions*, Series C, 1976, **98**, 550.
88. COMINI, G., DEL GIUDICE, S., LEWIS, R. W. and ZIENKIEWICZ, O. C. *Int. J. Numer. Method Eng.*, 1974, **8**, 613.
89. COMINI, G. and DEL GIUDICE, S. *J. Heat Transfer*, 1976, **98**, 543.
90. REBELLATO, L., DEL GIUDICE, S. and COMINI, G. *J. Food Sci.*, 1978, **43**, 239.
91. CARNAHAN, B., LUTHER, H. A. and WILKES, J. O. *Applied Numerical Methods*, John Wiley and Sons, Inc., New York, 1969.
92. TANDON, S. and BHOWMIK, S. R. *J. Food Sci.*, 1986, **51**(3), 709.
93. PEACEMAN, D. W. and RACHFORD, H. H. Jr. *J. Soc. Indust. Appl. Math.*, 1955, **3**(1), 28.
94. SMITH, G. D. *Numerical Solution of Partial Differential Equations, Finite Difference Methods*, 2nd edn, Oxford University Press, New York, 1978.
95. VICHNEVETSKY, R. and YADAV, R. K. *Analysis of a family of marching methods for parabolic equations*, Report NAM 192, Dept of Computer Science, Rutgers University, 1979.
96. VICHNEVETSKY, R. *IMACS Transactions*, 1979, **21**(2), 170.
97. CRANK, J. and NICOLSON, P. *Proc. Camb. Phil. Soc.*, 1947, **43**, 50.
98. DUFORT, E. C. and FRANKEL, S. P. *Math. Tables Aids Comput.*, 1953, **7**, 135.
99. HOLDSWORTH, S. D. *Food Manuf.*, 1968, **43**, 38.
100 FENNEMA, O. and POWRIE, W. D. *Adv. Food Res.*, 1964, **13**, 219.

Chapter 8

COMPUTER PREDICTION OF TEMPERATURES IN SOLID FOODS DURING HEATING OR COOLING

STUART THORNE

Department of Food and Nutritional Sciences, King's College, London, UK

SUMMARY

A micro-computer package is described which predicts temperature/time data for any point in a range of solids during heating or cooling by conduction from a uniform initial temperature in a medium at constant temperature. Mean temperatures can also be calculated. It is assumed that the solids are isotropic and homogeneous and that all the thermal and physical attributes of the solid and the heating medium are constant at all relevant temperatures. Although these conditions are rarely achieved during food processing—thermal and physical properties vary with temperature and mass transfer often accompanies heat transfer—the program predicts heating rates accurately in most cases. Solids in which temperatures can be calculated by the program are spheres, infinite cylinders, infinite slabs, infinite rectangular bars, finite cylinders and bricks. Other simple geometric shapes could easily be added.

Temperature/time data are plotted or tabulated on the computer screen or a printer and can be saved as a disk file for subsequent use. A directory of disk files previously saved can be printed and files can be selected from this directory for deletion, re-naming, plotting or sorting and tabulation. Several files can be plotted or tabulated simultaneously to facilitate investigation of variation in thermal or physical properties or heating or cooling conditions.

Examples of applications of the program are given.

NOTATION

a	Characteristic linear dimension	(m)
C_p	Specific heat	$(\text{kJ m}^{-3}\,^\circ\text{C}^{-1})$
F	Fourier Number $(k\,t)/(C_p\,r\,a^2)$	(Dimensionless)
h	Heat transfer coefficient	$(\text{kJ m}^{-2}\,\text{s}^{-1}\,^\circ\text{C}^{-1})$
j	Temperature $(T-T_m)/(T_i-T_m)$	(Dimensionless)
j_c	Centre temperature of a solid	(Dimensionless)
j_m	Mean temperature of a solid	(Dimensionless)
j_r	Temperature at position r	(Dimensionless)
k	Thermal conductivity	$(\text{kJ m}^{-1}\,^\circ\text{C}^{-1}\,\text{s}^{-1})$
n	An integer number $(n>0)$	(Dimensionless)
N	Biot number $(a\,h/k)$	(Dimensionless)
r	Distance from centre of solid	(m)
R	Position in solid (r/a)	(Dimensionless)
t	Time	(s)
T	Temperature	$(^\circ\text{C})$
T_i	Temperature at $t=0$	$(^\circ\text{C})$
T_m	Temperature of heating/cooling medium	$(^\circ\text{C})$
T_s	Temperature of heat transfer fluid	$(^\circ\text{C})$
V	Root of transcendental equation	(Dimensionless)
α	Thermal diffusivity $(k/a\,C_p)$	$(\text{m}^2\,\text{s}^{-1})$
ρ	Density	(kg m^{-3})

1. INTRODUCTION

Almost all foods are heated or cooled, intentionally or accidentally, during harvesting, transport, storage and processing and prediction of the rate of heating or cooling is vital for effective process design and optimisation.

The differential equation for heat conduction in a sphere, the surface of which is subjected to convective heat exchange[1,2] is:

$$\frac{\partial^2 j}{\partial R^2} + \frac{2}{R}\cdot\frac{\partial j}{\partial R} = \frac{\partial j}{\partial F} \tag{1}$$

for $0 \le R \le 1$ and $F \ge 0$.

If the sphere is initially at uniform temperature, the temperature of the heating or cooling medium is constant and the thermal and physical properties of the sphere are constant at all appropriate temperatures, then the solution to eqn (1) is:

$$j = \sum_{n=1}^{n=\infty} \left[\frac{2 \cdot N \cdot \sin(V_n \cdot R)}{R(V_n^2 + N^2 - N) \sin(V_n)} \cdot \exp(-V_n^2 F) \right] \tag{2}$$

where V_n is the nth root of the transcendental equation:

$$V \cot(V) = 1 - N \tag{3}$$

The mean temperature of a sphere is:

$$j_m = 3 \cdot \int_0^1 R^2 \cdot j(R, F) \cdot dR \tag{4}$$

$$j_m = \sum_{n=1}^{n=\infty} \left[\frac{6 \cdot N^2}{V_n^2(V_n^2 + N^2 - N)} \cdot \exp(-V_n^2 F) \right] \tag{5}$$

Similar equations (Appendix 1) can be derived for temperatures in infinite cylinders and infinite slabs. Temperatures in multi-dimensional solids can be expressed as the product of temperatures in appropriate one-dimensional solids. For example, a finite cylinder can be represented by the intersection at right angles of an infinite slab of thickness equal to the length of the finite cylinder and an infinite cylinder of the same diameter as the finite cylinder. The temperature at a point (x, y) can be expressed as;[2]

$$j_{(x, y)} = j_{(\infty - \text{slab}, x)} \times j_{(\infty - \text{cyl}, y)} \tag{6}$$

The traditional method for deriving numerical solutions to these equations has been the use of charts relating Fourier and Biot Numbers to dimensionless temperature.[2][5] However, such techniques lack precision, are slow and often apply only to a limited range of conditions. In particular, they usually apply only to a limited range of positions within the solid—typically predicting centre, surface and mean temperatures. The introduction of personal computers has facilitated the production of a computer package for the rapid solution of transient heat transfer equations.

This program assumes that foodstuffs are homogeneous and isotropic and the thermal and physical parameters are constant at all relevant temperatures. In practice, values do not vary greatly over the limited temperature ranges usually encountered in food processing operations and inhomogeneities and non-isotropic behaviour can be overcome by adoption of average values of thermal parameters. The limiting factor on the accuracy of prediction is likely to be the precision with which thermal and physical properties can be measured or predicted.

2. CAPABILITIES OF THE PROGRAM

The program produces numerical solutions to eqns (2), (3) and (5) and corresponding equations for infinite cylinders and infinite slabs. Solutions for finite cylinders, infinite rectangular bars and bricks are derived from temperatures in corresponding one-dimensional solids. When the program is started, the main menu offers a choice of options (Fig. 1). Each of these is described below.

HEATING & COOLING OF SOLIDS

Department of Food & Nutritional Sciences, King's College London

MAIN MENU

HIT APPROPRIATE FUNCTION KEY

INFORMATION	F1
CREATE TEMPERATURE/TIME FILES	F2
PRINT DIREC. OF TEMP/TIME FILES	F3
SELECT FILES FOR PRINTING	F4
DIRECTORY OF SELECTED FILES	F5
SORT & PRINT SELECTED FILES	F6
PLOT GRAPHS OF SELECTED FILES	F7
DELETE FILES	F8
RENAME FILES	F9
QUIT	F10

FIG. 1. Options available on the main menu.

2.1. Create Temperature/Time Files

Initially, the program assumes default values of all necessary physical and thermal attributes of the food and the heat transfer medium, but allows these to be changed as desired. Figure 2 is the screen display of the default

CURRENT PARAMETERS

Use up and down arrows on editing keys to select parameter to be changed and left arrow on editing keys to alter value. This is done by entering data, followed by ⟨RETURN⟩ for any variable except Shape, which will change itself when the left arrow is hit. Escape from this sequence by using "END" on the editing keys.

Shape of Solid	Sphere
Heat Transfer Coefficient	0·1
Specific Heat	4·18
Thermal Conductivity	4e–04
Density	1000
Minimum Time	0
Maximum Time	3000
Time Interval	300
Initial Temperature	0
Temp of Surroundings	100
Dimension 1 is Sphere of length	3e–02
Position on this dimension is	0

FIG. 2. Default values of thermal and physical parameters and calculation times. These can be altered as necessary.

values. No units are indicated, because all calculations use dimensionless groups. Shapes that can be handled are sphere, infinite circular cylinder, semi-infinite slab, infinite rectangular bar, finite cylinder and rectangular brick. When shape is changed, the display requests more or less dimensions and positions on these dimensions as appropriate. Position (r/a) has a value of zero for the centre to unity for the surface; values greater than unity compute mean temperatures. When this option is run, it calculates temperatures within the solid between the minimum and maximum times at specified time intervals. These are displayed graphically on the screen (Fig. 3) and can be tabulated on a printer. When the calculation is complete, the data can be saved to a disk file; such files are automatically given a name based on the system time and date. The previous set of choices of thermal and physical parameters are re-displayed and can be altered before calculating another set of temperature/time data.

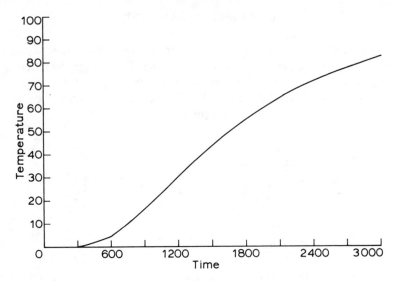

FIG. 3. Screen output when the program is run with default values. These data can be saved as a disk file.

2.2. Print Directory of Temperature/Time Files

This prints, either to the screen or to the printer, a list of all temperature/time disk files which have been saved, together with all information about the thermal and physical parameters used.

2.3. Select Files for Printing

One of the features of the program is that it can print graphs or tables to the screen or printer of two or more disk files of temperature/time data. This, for example, allows easy comparison of the effect on the rate of heating of altering the heat transfer coefficient or the physical properties of a system. Before this can be done, files to be printed have to be selected from the complete list of files available. This option displays, on the screen, file names and thermal and physical parameters appropriate to the file (Fig. 4). Because the complete list of parameters cannot be displayed simultaneously, the list can be scrolled up and down to display all the files and from side to side to view all the thermal and physical parameters of the files.

2.4. Directory of Selected Files

This prints a directory of files selected for printing by the previous option and in the same style.

SELECT FILES FOR PRINTING

Use ↑ & ↓ edit keys to choose files and "ENTER" to add to print list, Left & Right edit keys display file values. "END" escapes sequence.

FILENAME	H	C	Shape	Rho	K	Tmin	Tmax
26FB1725.@55	1e–02	3·45	Sphere	804	3·8e–040		5000
26FB1726.@36	2e–02	3·45	Sphere	804	3·8e–040		5000
26FB1727.@18	5e–02	3·45	Sphere	804	3·8e–040		5000
26FB1728.@04	0·1	3·45	Sphere	804	3·8e–040		5000
26FB1728.@59	0·2	3·45	Sphere	804	3·8e–040		5000
26FB1729.@45	0·5	3·45	Sphere	804	3·8e–040		5000

FIG. 4. Directory of temperature/time data files saved to disk. Complete information about these can be viewed by scrolling.

2.5. Sort and Print Selected Files

Files selected for printing can be tabulated on the printer. Times are sorted so that all times encountered in any of the selected files are listed in order and temperatures for each file are printed in subsequent columns if they were calculated. Temperatures can be printed either as dimensionless temperature (j) or as real temperatures (e.g. °C). Table 1 is an example of output from this routine, showing temperatures in a sphere between the centre and the surface at intervals of $R = 0.2$, using the default values of Fig. 2. The file for the surface calculated temperatures at 150 s intervals between 0 and 2000 s, whereas the file for the surface calculated temperatures at 150 s; the others at 300 s intervals between 0 and 3000 s.

TABLE 1
TYPICAL PRINTOUT OF TEMPERATURE/TIME DATA

Time (s)	SPH0	SPH2	SPH4	SPH6	SPH8	SPH10
0	0	0	0	0	0	0
150	—	—	—	—	—	60·00
300	0·25	0·346	2·21	9·286	31·31	70·70
450	—	—	—	—	—	76·76
600	4·52	6·48	13·57	28·38	52·01	80·78
750	—	—	—	—	—	83·74
900	16.55	19.53	28·67	44·02	64·31	86·05
1050	—	—	—	—	—	87·90
1200	31·02	33·99	42·66	56·16	72·67	89·43
1350	—	—	—	—	—	90·72
1500	44·47	47·05	54·44	65·59	78·77	91·83
1650	—	—	—	—	—	92·79
1800	55·84	57·96	63·99	72·95	83·40	93·62
1950	—	—	—	—	—	93·36
2100	65·08	66·78	71·60	78·73	86·98	—
2400	72·46	73·81	77·63	83·27	89·77	—
2700	78·31	79·38	82·39	86·84	91·95	—
3000	82·93	83·77	86·14	89·64	93·67	—

2.6. Plot Graphs of Selected Files

This routine uses the same data as in Section 2.5, but plots it as a graph on the screen. Legends can then be typed on the graph before it is transferred to the printer. Figure 5 is a typical graph printed by the program.

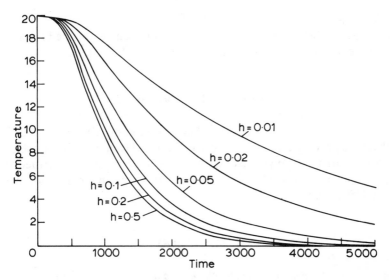

FIG. 5. Printer output of centre temperatures of apples for $h = 0.01$ to $h = 0.5$ kJ m^{-2} s^{-1} °C.

2.7. Delete Files

Files are displayed on the screen in a format similar to that described in Section 2.3 and can be selected for deletion from the directory.

2.8. Rename Files

When temperature/time files are created, they are automatically named by date and time from the system clock; a file created on July 26th at 2 h 59 min and 56 s would have the name '26JL259.@56', which is hardly a useful indication of its contents! This option allows file names to be changed to permit identification. Files are displayed as described in Section 2.3 and can be selected for re-naming. The selected list is then displayed and the user is prompted for a new name, with an option to leave the name unchanged.

Renaming or deleting files that are contained on the list selected for printing (Section 2.3) deletes that list also.

2.9. Quit

When the program is started, all existing data files are copied to a RAM disk for rapid execution. The quit option ensures that all files are closed, unwanted files are deleted and new files copied to the default drive before returning to the operating system.

2.10. Help

Information about the program and its options can be obtained from the main menu with the 'Help' option.

3. SYSTEM REQUIREMENTS

The program operates under MSDOS[†] and requires a minimum of 128 Kbytes of RAM and a single disk drive. For rapid handling of files, a RAM disk of at least 256 Kbytes is desirable. The program has been written for operation on a RM Nimbus[‡] micro-computer, but can be adapted for most MSDOS machines.

4. SOFTWARE

Those parts of the main program which produce numerical solutions to heat transfer equations are listed in Appendix 2. Numbers in square brackets refer to line numbers in this listing. The program is written in RM Basic.[2]

Equation (2) or eqn (5) for spheres or either of the corresponding equations for other shapes is solved by adding terms [Sphere 3230–3290, Infinite Cylinder 3300–3360, Infinite Slab 3370–3430] until the change in the sum resulting from inclusion of the next term is less than a pre-determined value [Sphere 3670–3740, Infinite Cylinder 3830–3900, Infinite Slab 3750–3820].

Roots of the transcendental eqn (3) and corresponding equations for other one-dimensional solids are derived by searching iteratively between the known limits for each root [Sphere 2500–2630, Infinite Slab 2640–2730, Infinite Cylinder 2740–2870]. These are calculated the first time they are needed and are then stored in an array and used when necessary. As the greatest number of roots are usually needed for short times, the program does not normally calculate roots except at the start of a run. If program parameters are changed, roots are only recalculated if a, h, k or shape of the solid is changed. Solutions for temperatures in cylinders require Bessel functions J_0 and J_1; these are not standard functions and are calculated [J_0 2880–2990, J_1 3000–3110]. Two algorithms are used for each of the Bessel functions. For arguments < 10, Bessel functions are calculated as

---

[†]Microsoft Ltd, Piper House, Hatch Lane, Windsor, Berks SL4 3QJ, UK.
[‡]Research Machines Ltd, Mill Street, Oxford OX2 OBW, UK.

a fundamental infinite series,[6] terms being added until the difference between (n) and $(n+1)$ terms is negligibly small [2900–2940; 3020–3060]. For arguments >10, an approximation, invalid for small arguments, is used[6] because a great number of terms of the infinite series are needed for adequate precision [2960–2970; 3080–3090]. The limits of the transcendental equation for an infinite cylinder are the zero points of (J_1/J_0),[7] which are stored as data statements [4210–4240].

For multi-dimensional solutions, terms of eqns (2) or (5) or corresponding equations for other shapes are calculated for each dimension. Temperatures for multi-dimensional solids are calculated from the products of the dimensionless temperatures of each dimension, as exemplified by eqn (6).

5. EXAMPLES OF APPLICATIONS

5.1. Effect of Aspect Ratio on the Rate of Heating of Canned Foods

Rates of heating of cylindrical cans of potato purée of 540 ml capacity and lengths of between 4 cm and 20 cm in steam at 125°C, including a standard 300×411 can (length $= 11.9$ cm, diameter $= 7.62$ cm, capacity $= 540$ ml) were computed. Thermal diffusivity of the purée was determined by the method of Dickerson.[8] Figure 6 shows the effect of can length and diameter

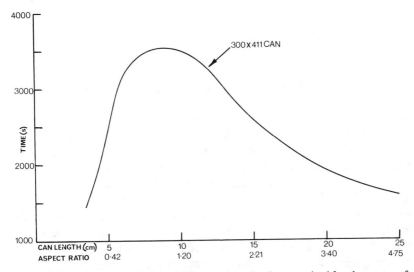

FIG. 6. Effect of aspect ratio (length/diameter) on the time required for the centre of a cylindrical can to achieve $j = 0.5$ ('half-cooling time').

on predicted rates of heating, expressed as the 'half-heating time'; the time required to achieve a centre temperature of $j_C = 0.5$ (77·5°C). This demonstrates that the standard 300×411 can is not the ideal shape for rapid heating. However, increasing the diameter would reduce the mechanical rigidity of cans and reducing it would produce an aesthetically undesirable shape. Cans (11·9 cm long × 7·62 cm diameter) of the potato purée were heated in steam, the centre temperature being measured with fine wire thermocouples.[10] Throughout heating, the predicted and observed temperatures differed by less than 1°C (Table 2).

TABLE 2

PREDICTED AND OBSERVED TEMPERATURES IN CANS OF POTATO PURÉE

Time (s)	Observed temperature (°C)	Predicted temperature (°C)
0	30·1	30·0
1000	33·5	32·8
1500	42·0	42·5
2000	54·9	54·2
2500	66·3	65·7
3000	76·6	76·0
3500	85·7	84·8
4000	93·1	92·3
4500	99·0	98·4
5000	104·0	103·4

5.2. Effect of Overall Heat Transfer Rate on the Rate of Cooling of Apple Fruit

Data for the thermal and physical properties of apples were extracted from the literature[9,11-14] ($k = 3.8 \times 10^{-4}$ kJ m^{-1} s^{-1} °C^{-1}, $c = 3.45$ kJ kg^{-1} °C^{-1}, $\rho = 804$ kg m^{-3}). These were used to predict the rate of cooling of spherical apples of diameter 6·0 cm from an initial temperature of 20°C in cooling media at 1°C with overall heat transfer coefficients between 0·01 and 0·5 kJ m^{-2} s^{-1} °C^{-1}. The results of this are given in Fig. 5. Actual apples with fine wire thermocouples mounted at their centres were cooled in air ($h =$ kJ m^{-2} s^{-1} °C^{-1}) and water ($h = 0.2$ kJ m^{-2} s^{-1} °C^{-1}).[9] Temperatures throughout cooling were within ± 1°C of those predicted by the program. Dimensionless temperatures for the predicted and observed data were calculated; their slopes were found not to be significantly different. This was in spite of the imperfect sphericity of the fruit and the uncertainty of locating their centres.

5.3. Temperature Distribution in an Infinite Slab during Heating

An aluminium container 15 cm × 15 cm × 7 cm was fitted with fine wire thermocouple at intervals of 0·35 cm between its centre and surface. It was filled with the potato purée described in Section 5.1 and heated from an initial temperature of 2°C in water at 100°C. The overall heat transfer coefficient from the water to an aluminium sphere was measured and assumed to be similar to that to the metal container. Calculated temperature distributions in the slab are given in Fig. 7. Again, observed temperatures were within ±1°C of calculated temperatures in all cases.

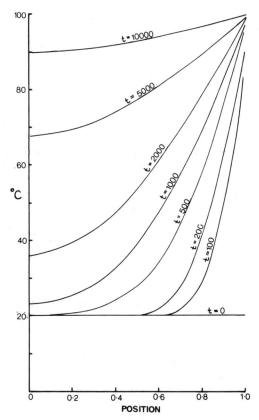

FIG. 7. Temperature distribution in a slab during heating (0, centre; 1, surface).

REFERENCES

1. CARSLAW, N. S. and JAEGER, J. C. *The Conduction of Heat in Solids*, Clarendon Press, Oxford, 1947.
2. DALGLEISH, N. and EDE, A. J. *Charts for determining Centre, Surface and Mean Temperatures in Regular Geometric Solids during Heating and Cooling*, National Engineering Laboratory Report No. 192, Ministry of Technology, East Kilbride, 1965.
3. MEFFERT, H. F. Th. *A New Chart for the solution of Transient Heat Transfer Problems*, Institut International du Froid, Annexe 1970, IIF, Paris, 1971.
4. HEISLER, M. P. Temperature Charts for Induction and Constant Temperature Heating. *Trans. Amer. Soc. Mech. Engrs*, 1947, **69**(3), 227–36.
5. NEWMAN, A. B. Heating and Cooling Rectangular and Cylindrical Solids. *Industr, Engng Chem.*, 1936, **28**(5), 545–8.
6. ABRAMOVITCH, M. and STEGUN, A. *Handbook of Mathematical Functions*, New York, 1965.
7. FLÜGGE, W. *Four Place Tables of Transcendental Equations*, Pergamon, London, 1954.
8. DICKERSON, R. W. An apparatus for the Measurement of Thermal Diffusivity of Foods. *Food Technol.*, 1965, **19**, 880.
9. GORDON, C. MCD. Computer-aided determinations of Thermal Properties of Foods, Ph.D. thesis, 1986, University of London.
10. COWELL, N. D., EVANS, H. L., HICKS, E. W. and MELLOR, J. D. *Food Technol.*, 1959, **13**, 425.
11. GANE, R. *Thermal Conductivity of the Tissue of Fruits*, Rep. Food Invest. Board, London, 1936, p. 211.
12. MOTAWI, H. Measurements of Cooling Rates of Fruits and Vegetables, M.Sc. thesis, 1962, Michigan State University.
13. FRECHETTE, R. J. and ZANRADNIK, R. J. Thermal Properties of the McIntosh Apple. *ASAE Transactions*, 1968, **11**(1).
14. BENNETT, A. H., CHANCE, W. G. and CUBBEDGE, R. H. Heat Transfer properties and characteristics of Appalachian area Red Delicious Apples. *ASHRAE Transactions*, 1969, **75**(2), 133.

APPENDIX 1
TEMPERATURES IN INFINITE CYLINDERS AND SLABS

1. Infinite Cylinders

The temperature at a point R is;

$$j = \sum_{n=1}^{n=\infty} \left[\frac{2 \cdot N \cdot J_0(V_n \cdot R)}{(V_n^2 + N^2)J_0(V_n)} \cdot \exp(-V_n^2 F) \right] \tag{7}$$

where J_0 is a Bessel function of the first kind and V_n is the nth root of the transcendental equation:

$$N/V = J_1(V)/J_0(V) \tag{8}$$

The mean temperature in an infinite cylinder is:

$$j_m = 2 \cdot \int_0^1 R \cdot j(R, F) \cdot dR \tag{9}$$

$$j_m = \sum_{n=1}^{n=\infty} \left[\frac{4 \cdot N^2}{V_n^2(V_n^2 + N^2)} \cdot \exp(-V_n^2 F) \right] \tag{10}$$

2. Infinite Slab

The temperature at a point R in an infinite slab is:

$$j = \sum_{n=1}^{n=\infty} \left[\frac{2 \cdot N \cdot \cos(V_n \cdot R)}{(V_n^2 + N^2 + N)\, \cos(V_n)} \cdot \exp(-V_n^2 F) \right] \tag{11}$$

where V_n is the nth root of the transcendental equation;

$$V \cdot \tan(V) = N \tag{12}$$

The mean temperature is:

$$j_m = \frac{1}{2} \int_0^1 j(R, F) \cdot dR \tag{13}$$

$$j_m = \sum_{n=1}^{n=\infty} \left[\frac{2 \cdot N^2}{V_n^2(V_n^2 + N^2 + N)} \cdot \exp(-V_n^2 F) \right] \tag{14}$$

APPENDIX 2
LISTING OF SUB-ROUTINES FROM THE MAIN PROGRAM

1. Calculate n^{th} Root (V_n) of Equation (3)

```
2500    FUNCTION SphereRoot(N, n)
2510        IF N = 1 THEN Cent := n + 0·5 : GOTO 2610
2520        IF N < =1 THEN Ls := 0 ELSE Ls := 0·5
2530        Lower := n + Ls : Cent := Lower + 0·25 : Upper := Cent + 0·25
2540        V := 0
2550        REPEAT
```

```
2560        V := (1 − (((Cent * PI)/(TAN(Cent * PI)))))
2570        IF N < V THEN Upper := Cent ELSE Lower := Cent
2580        Cent := (Upper + Lower)/2
2590        IF ABS(Upper − Lower) < 1e-04 THEN V := N
2600    UNTIL ABS(V − N) < N * 1e–04
2610    V := Cent * PI
2620    RESULT V
2630 ENDFUN
```

2. Calculate n^{th} Root (V_n) of Equation (12)

```
2640    FUNCTION SlabRoot(N, n)
2650        Lower := n : Cent := Lower + 0·25 : Upper := Cent + 0·25 : V := 0
2660        REPEAT
2670            4 := Cent * PI * TAN(Cent * PI)
2680            IF N < V THEN Upper := Cent ELSE Lower := Cent
2690            Cent := (Upper + Lower)/2
2700        UNTIL ABS(V − N) < N * 1e–04
2710        V := Cent * PI
2720        RESULT V
2730    ENDFUN
```

3. Calculate n^{th} Root (V_n) of Equation (8)

```
2740    FUNCTION CylinderRoot(N, n)
2750        RESTORE 4210
2760        FOR Coop% := 0 TO n : READ Lower, Upper : NEXT Coop%
2770        Cent := (Lower + Upper)/2 : V := 0
2780        REPEAT
2790            First := J1(Cent)
2800            Second := J0(Cent)
2810            Bessy := First / Second
2820            V := Cent * Bessy
2830            IF N < V THEN Upper := Cent ELSE Lower := Cent
2840            Cent := (Lower + Upper)/2
2850        UNTIL ABS(V − N) < 1e–03 * N
2860        RESULT Cent
2870    ENDFUN
```

4. Calculate $J_0(X)$

```
2880    FUNCTION J_0(X)
2890        IF X > = 10 THEN 2960
```

```
2900    Nn% := INT(X + 7): Bes := 1 : Cc := 1
2910    FOR Aoop% := 2 TO (2 * Nn% − 2) STEP 2
2920      Cc := Cc * (Aoop%^2)
2930      IF INT(Aoop%/4) = Aoop%/4 THEN Bes := + ((X^
            Aoop%)/Cc) ELSE Bes := Bes − ((X^Aoop%)/Cc)
2940    NEXT Aoop%
2950    GOTO 2980
2960    PO := 1 − (7·3125e–02/(X^2)) + (2·604e–03/(X^4)): Q0 := −
            (0·125/X) + (2·6367e–02/(X^3))
2970    Bes := SQRT(2/(PI*X)) * ((P0 * COS(X − (PI/4))) − (Q0 *
            SIN(X − (PI/4))))
2980    RESULT Bes
2990    ENDFUN
```

5. Calculate J_1 (X)

```
3000    FUNCTION J_1(X)
3010      IF X > = 10 THEN 3080
3020      Nn% := INT(X + 7): Bes := X/2 : Cc := 1
3030      FOR Boop% := 2 TO (2 * Nn% − 2) STEP 2
3040        Cc := Cc * (Boop%^2)
3050        IF INT(Boop%/4) = Boop%/4 THEN Bes := Bes + (X^
              (Boop% + 1))/(Cc * (Boop% + 2)) ELSE Bes := Bes − (X^
              (Boop% + 1))/(Cc * (Boop% + 2))
3060      NEXT Boop%
3070      GOTO 3100
3080      P1 := 1 + (0·117188/(X^2)) − (0·1441955/(X^4)) + (0·675926/
              (X^6)): Q1 := (0·375/X) − (0·102536/(X^3)) + (0·2775764/
              (X^5))
3090      Bes := SQRT(2/(PI * X)) * ((P1 * COS(X − (3 * PI/4))) −
              − (Q1 * SIN(X − (3 * PI/4))))
3100    RESULT Bes
3110    ENDFUN
```

6. Sum of Terms of Equation (2) and Equation (5)

```
3230    FUNCTION SphereSum(V, R, N, F)
3235      IF (V^2 * F) > 80 THEN Term := 0 : GOTO 3280
3240      IF R = 0 THEN Term := (2 * N * V)/((V^2 + N^2 − N) *
            SIN(V)) * EXP(−(V^2 * F))
3250      IF R > 0 AND R < 1 THEN Term := (2 * N * SIN(V * R))/(R *
            (V^2 + N^2 − N) * SIN(V)) * EXP(−(V^2 * F))
```

3260   IF R = 1 THEN Term := $(2*N)/(R*(V^2 + N^2 - N))*$
          $EXP(-(V^2*F))$
3270   IF R > 1 THEN Term := $(6*N^2)/(V^2*(V^2 + N^2 - N))$
          $*EXP(-(V^2*F))$
3280   RESULT Term
3290   ENDFUN

7. Sum of Terms of Equation (7) and Equation (10)

3300   FUNCTION CylinderSum(V, R, N, F)
3305   IF $(V^2*F) > 80$ THEN Term := 0 GOTO 3350
3310   IF R = 0 THEN Term := $(2*N)/((V^2 + N^2)*J0(V))*$
          $EXP(-(V^2*F))$
3320   IF R > 0 AND R < 1 THEN Term:=$(2*N*J0(V*R))/((V^2$
          $+N^2)*J0(V))*EXP(-(V^2*F))$
3330   IF R = 1 THEN Term := $(2*N)/(V^2 + N^2)*EXP(-$
          $(V^2*F))$
3340   IF R > 1 THEN Term := $(4*N^2)/(V^2*(V^2 + N^2))*$
          $EXP(-(V^2*F))$
3350   RESULT Term
3360   ENDFUN

8. Sum of Terms of Equation (11) and Equation (14)

3370   FUNCTION SlabSum(V, R, N, F)
3375   IF $(V^2*F) > 80$ THEN Term := 0:GOTO 3420
3380   IF R=0 THEN Term := $(2*N)/((N^2+N+V^2)*COS(V))$
          $*EXP(-(V^2*F))$
3390   IF R > 0 AND R < 1 THEN Term := $(2*N*COS(V*R))/$
          $((N^2 + N + V^2)*COS(V))*EXP(-(V^2*F))$
3400   IF R = 1 THEN Term := $(2*N)/(N^2 + N + V^2)*$
          $EXP(-(V^2*F))$
3410   IF R > 1 THEN Term := $(2*N^2)/((V^2 + N^2 + N)*V^2)$
          $*EXP(-(V^2*F))$
3420   RESULT Term
3430   ENDFUN

9. Calculation of Temperatures

Variables used are;

Tmin   Minimum Time
Tmax   Maximum Time
Tint   Time interval

A (X)	Size of solid in dimension (X)
Shape$ (X)	Shape of solid in dimension (X)
Hardcopy%	If "true" sends output to line printer
Thetai	Initial temperature
Thetas	Temperature of surroundings
Xbot,Xrange,Ybot,Yrange	Scales for graph plotting

```
3440 PROCEDURE CalculateTemperature
3450    GLOBAL Tmin, Tmax, Tint, A( ), h, k, Rho, c, Shape$( ), V( ),
        R( ), n, Hardcopy%, Thetai, Thetas, Xbot, Xrange, Ybot,
        Yrange
3460    Xprev := Tmin : Yprev := Thetai
3470    FOR Ptime := Tmin TO Tmax STEP Tint
3480       Thetar := 1 : IF Ptime = 0 THEN 3550
3490       FOR Loop% := 1 TO n
3500          A$ := GET$(0) : IF A$ = CHR$(27) THEN LEAVE
3510          N := (Loop%) * H / K : F := (K * Ptime) / (Rho * C *
              A(Loop%)^2)
3520          IF Shape$(Loop%) = "Sphere" THEN GOSUB 3670
              ELSE IF Shape$(Loop%) = "Slab" THEN GOSUB 3750
              ELSE GOSUB 3830
3530          Thetar := Thetar * Sum
3540       NEXT Loop%
3545       IF Thetar > 1 THEN Thetar = 1
3550       Thetac := Thetar * (Thetai − Thetas) + Thetas
3560       Thetar := INT(Thetar * 1000) / 1000 : Thetac := INT(Thetac *
           1000) / 1000
3570       IF Hardcopy% = 1 THEN PRINT £2 TAB (10); Ptime; TAB
           (35); Thetar; TAB (60); Thetac
3580       PRINT ~ 2 TAB (10); "Time = "; Ptime; TAB (30); "Thetar = ";
           Thetar; TAB (50); "Thetac = "; Thetac
3590       PRINT £20, Ptime; CHR$(10); Thetar; CHR$(10); Thetac
3600       Xprev := ((Xprev − Xbot) / Xrange) * 550 + 65 : Yprev :=
           ((Yprev − Ybot) / Yrange) * 180 + 55
3610       Xnew := ((Ptime − Xbot) / Xrange) * 550 + 65 : Ynew :=
           ((Thetac − Ybot) / Yrange) * 180 + 55
3620       LINE Xprev, Yprev; Xnew, Ynew BRUSH 3
3630       Xprev := Ptime : Yprev := Thetac
3640    NEXT Ptime
3650 LEAVE
3660 STOP
```

```
3670  Loop2% := 1 : Sum := 0
3680  REPEAT
3690     Prevsum := Sum
3700     IF V(Loop%, Loop2%) = 0 THEN V(Loop%, Loop2%) :=
         Sroot(N, Loop2% − 1)
3710     Sum := Sum + Spsum(V(Loop%, Loop2%), R(Loop%), N,
         F)
3720     Loop2% := Loop2% + 1
3730  UNTIL ABS(Prevsum − Sum) < 1e−02 OR Loop2% = 5
3740  RETURN
3750  Loop2% := 1 : Sum := 0
3760  REPEAT
3770     Prevsum := Sum
3780     IF V(Loop%, Loop2%) = 0 THEN V(Loop%, Loop2%) :=
         Slroot(N, Loop2% + 1)
3790     Sum := Sum + Slsum(V(Loop%, Loop2%), R(Loop%) N, F)
3800     Loop2% := Loop2% + 1
3810  UNTIL ABS(Prevsum − Sum) < 1e−02 OR Loop2% > 5
3820  RETURN
3830  Loop2% := 1 : Sum := 0
3840  REPEAT
3850     Prevsum := Sum
3860     IF V(Loop%, Loop2%) = 0 THEN V(Loop%, Loop2%) :=
         Cyroot(N, Loop2% − 1)
3870     Sum := Sum + Cysum(V(Loop%, Loop2%), R(Loop%), N,
         F)
3880     Loop2% := Loop2% + 1
3890  UNTIL ABS(Prevsum − Sum) < 1e−02 OR Loop2% = 5
3900  RETURN
3910  ENDPROC
```

10. Zero Points of (J_1/J_0)

```
4210  DATA 0, 2·40482555, 3·83170595, 5·52007845, 7·0155869,
      8·6535261, 10·1734679
4220  DATA 11·79156,13·32369,14·93093,16·47063,18·07107, 19·61586
4230  DATA 21·21164,22·76008,24·35247,25·90367,27·49348, 29·04683
4240  DATA 30·63461,32·18968,33·77582,35·33233,36·9171,38·47477,
      40·05843
```

INDEX